A SHORT HISTORY

OF

BALLOONS AND FLYING MACHINES.

A SHORT HISTORY

OF

BALLOONS

AND

FLYING MACHINES.

EDITED BY

LORD MONTAGU.

With a Chapter by

MAJOR B. BADEN-POWELL.

LONDON:

"THE CAR ILLUSTRATED,"

168, PICCADILLY, W.

1907.

FLYING MACHINES.

"O that I had wings like a dove! for then would I flee away and be at rest."—*King David.*

"'Tis easier sport than the balloon."—*Heywood.*

T HE earliest mention of flying seems to be that of Dædalus, a figure in Greek mythology who personified the beginning of the Arts of Sculpture and Architecture. He was of the old Athenian royal race of the Erechtheidæ. Having killed his nephew and pupil in envy at his growing skill, he had to flee to Crete, where he made the well-known cow for Queen Pasiphaë.

The romantic tradition concerning Dædalus is as follows:—Having committed this great crime, he fled from Athens to Crete, taking with him his son Icarus. He there constructed for Minos, King of that island, the famous labyrinth, with which every one is familiar; but, having incurred the King's displeasure, he was himself confined therein. In order to effect his escape he made wings of feathers and wax for himself

A

and his son, and with these attempted to fly away; but Icarus soared so high that the sun melted the wax by which his wings were fastened, and he fell into that part of the sea which, by way of testimony, bore his name for hundreds of years afterwards.

Dædalus, however, more careful, arrived safely in Sicily.

1260.—At this date Roger Bacon is the first English philosopher who asserts the existence of a machine for flying; but he says, "Not that he himself had seen it, or was acquainted with any person who had done so, but he knew an ingenious person who had contrived one."

1338.—At this date lived Froissart, the great historian; he speaks of an apprentice of Valenciennes who made himself a pair of wings six feet and a half high, and requested the Count of Flanders to be allowed to try them in his presence. Of course the Count assented, and, in some curiosity, came out with his whole Court to see the sight; but the performance was exceedingly meagre.

After tying his wings to his shoulders, the apprentice was taken up to an embrasure that overlooked the castle drawbridge, and he was just on the point of leaping forward, when the Countess became nervous, and ordered him to an embrasure that crowned the moat, so that in case of accident he should only get a ducking. This turned out to be a good precaution, for, upon jumping from the battlements, the venturesome lad tumbled head foremost into the water.

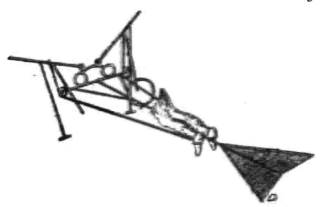

THE INVENTION OF LEONARD DE VINCI, ABOUT 1500.

1510.—The first *historical* flying experiment was made this year in Scotland by an Italian Friar, whom King James IV. had made Prior of Tongland.

This man was a great favourite, from his supposed successes in alchemy; in fact, he was said to be in league with "*Auld Hornie*"; and thinking he had discovered a method of flying, the Prior appointed a day for a flight, and invited the King and his Court to witness the feat. At the appointed time the Italian, with an enormous pair of wings, ascended one of the battlements of Stirling Castle, and in the presence of James and his Court spread his wings and vaulted into the air.

Unfortunately, the experiment was a complete failure; he came down anyhow, and tumbled on to a manure heap, which saved his neck; but he broke his thigh bone.

A 2

He said his failure was owing to the wings of his machine containing common feathers from common poultry, instead of being all pure from eagles and other noble birds.

1617.—This year a monk of Tubingen made himself wings of parchment, and leaped with them into the air from a high tower: he fell to the ground and was killed.

Previous to this, Fielder, rector of the School at Tubingen, gave a lecture on the art of flying, which the monk unfortunately illustrated with the above fatal result.

1640.—At this period a Frenchman named Cyrano de Bergerac wrote *The History of the States and Empires of the Moon and Sun*, and, speaking of his first imaginary voyage to the moon, he writes:—

"He filled with smoke two large vases, which he sealed hermetically and fastened under his wings; hereupon the smoke, which had a specific tendency to rise, but which was unable to penetrate the metal, pushed the vases upwards so that they rose into the clouds, carrying with them this great man.

"And he, when he had reached to 25 feet above the surface of the moon, untied the vases he had got as wings around his shoulders and allowed himself to fall.

"The height was great, but he wore a long and ample gown, into the folds of which the wind engulfed itself, and thus bore him softly and slowly to earth" (of the moon).

Now it is not only a remarkable, but a very curious coincidence, that this was written before the Montgolfiers were born, and about 143 years before they started their first balloon which was in 1783, and which ascended by means of *smoke*.

1645.—Cyrano de Bergerac wrote another work, the title whereof was *The Comical History of the Kingdom of the Sun and the Moon*.

1660.—This year Francis Lana, a Jesuit Priest, proposed to make hollow spheres of copper, which being exhausted of air, would float in our ordinary atmosphere.

1672.—John Wilkins, Bishop of Chester, studied mathematics and mechanical philosophy. He wrote a curious treatise, viz., *Discovery of a New World*, 1638, which gravely discussed the possibility of communicating by a *flying machine* with the moon and its supposed inhabitants, and he stated that it would be possible to make a journey there, if he could be conveyed for a starting point to some place beyond the reach of the earth's attraction.

He also published in 1691 (fourth edition) a work called *Mathematical Magick: or the Wonders that may be performed by Mechanical Geometry*, in two books:—

    1. "Archimedes, or Mechanical Powers."

    2. "Dædalus, or Mechanical Motion."

In the latter—

    Chap. 6: "Of the volant automata. Archytos his dove, and Regiomoutanus his eagle."

Chap. 7 : "Concerning the art of flying
the several ways hereby this hath been
or may be attempted."
Chap. 8 : "A resolution of the two chief
difficulties that seem to oppose the possibility of a flying chariot."

1678.—MODEL BY BESNIER.

1709.—No. 56 of *The Evening Post,* a newspaper
published in the reign of Queen Anne, and
dated 20-22 December, 1709, sets out a description of a flying ship, the invention of a Brazilian
priest, Bartholomew Laurent, and brought under
the notice of the King of Portugal. He says :—

"That he has found out an invention by
the help of which one may more speedily travel
through the air than any other way, either by
sea or land, so that one may go 200 miles in
twenty-four hours. Merchants may have their
merchandize, and send letters and packets more
conveniently. Places besieged may be supplied

with necessaries and succours: moreover, we may transport out of such places what we please and the enemy cannot hinder it.*

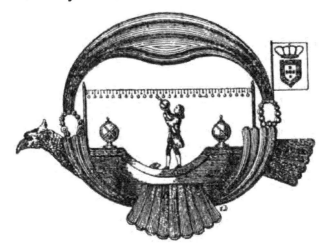

1709.—BARTHOLOMEW LAURENT.

1736.—*The Gentleman's Magazine* for this year records that " on the evening of the 1st October, during the performance of an entertainment called " *Dr. Faustus* " at Covent Garden Theatre, one James Todd, who represented the " Miller's Man," fell from the upper stage in a flying machine by the breaking of the wires. He fractured his skull and died miserably: three others were much hurt, but recovered."

---

* This actually came true, as witness the siege of Paris and the Balloon Post.

1742.—In this year the Marquis de Bacqueville announced that he would fly with wings from the top of his own house on the Quai des Theatins to the Gardens of the Tuileries.

He actually accomplished half the distance, when, being exhausted with his efforts, the wings no longer beat the air, and he came

MARQUIS DE BACQUEVILLE.

down into the Seine, and would have escaped unhurt but that he fell against one of the floating machines of the Parisian laundresses and thereby fractured his leg.

1753.—The earliest advertisement having reference to balloons is that in the *Public Advertiser* of September 24th, 1753:—

"*MARYBONE GARDENS.

"The Musical Entertainments at this place will end this Evening.

"The doors to be opened at 5 o'clock, the Music to begin exactly at six, and the Fireworks at nine. 2 Sky Rockets, 2 Air Balloons, 2 Balloons, and a large Balloon Wheel, which throws out of eight boxes Stars and Serpents.

"Admittance One Shilling."

° So spelt.

1758.—About this time Dr. Johnson wrote his *Rasselas.* In Chapter VI. is a dissertation on the art of flying. He says:—

"Among the artists that had been allured into the Happy Valley to labour for the accommodation and pleasure of its inhabitants was a man eminent for his knowledge of the mechanic powers, who had contrived many engines, both for use and recreation.

"One day he was found busy in building a sailing chariot. 'Sir,' said the master, 'You have seen but a small part of what the mechanic arts can perform. I have long been of opinion that instead of the tardy conveyance of ships and chariots man might use the swifter migration of wings, that the fields of air are open to knowledge, and that only ignorance and idleness need crawl upon the ground.'

"'The labour of rising from the ground will be great,' said the artist, 'as we see it in the heavier domestic fowls; but as we mount higher, the earth's attraction and the body's gravity will be gradually diminished, till we arrive at a region where man shall float in the air without any tendency to fall.'

"'Nothing,' replied the artist, 'will ever be attempted if all possible objections must be first overcome. If you will favour my project, I will by the first flight at my own hazard convince you. I have considered the structure of all volant animals, and find the folding continuity of the bat's wings most easily accommodated to

the human form.   Upon this model I will begin
my task to-morrow, and in a year I expect to
tower into the air beyond the malice and pur-
suit of man.'

"In a year the wings were finished, and on
the morning appointed the maker appeared,
furnished for flight, on a little promontory: he
waved his pinions awhile to gather air, then
leaped from his stand, and in an instant
dropped into the lake.

"His wings, which were of no use in the air,
sustained him in the water, and the Prince drew
him to land half dead with terror and vexation."

1784.—James Tytler, surgeon, chemist, aëronaut, litterateur, and poet, the editor of the
second edition of the *Encyclopædia Britannica*,

1784. —APPARATUS INVENTED BY GERARD.

after failing in two attempts, ascended from Comely Gardens in a fire balloon, stove and all, to a height of 350 feet.

1817.—In the foreign journals of this year there was the following announcement :—

Flying Machine. "A country clergyman in Lower Saxony has been so happy as to succeed in accomplishing the invention of an air-ship.

"The machine is built of light wood; it is made to float in the air, chiefly by means of the constant action of a large pair of bellows of a peculiar construction, which occupies in the front the position of the lungs, and the neck of a bird on the wing; the wings on both sides are directed by cords.

"The height to which a farmer's boy about ten or twelve years old, whom the inventor had instructed in the management of it, had hitherto ascended with it, is not considerable, because his attention has been more directed to give a progressive than an ascending motion to his machine."

All the above records are facts so far as relates to flying machines. But it is worthy of note that a strange theory has been put forward, that during life, the quills of birds, as well as their hollow bones, are filled with hydrogen.

"Flying animals," says a writer in *All the Year Round* for March 7th, 1868, "are built to hold gases everywhere—in their bones, their bodies, their skins; and their blood is several degrees warmer than the blood of walking or running animals: their gases are probably several degrees lighter.

"Azote or hydrogen, or whatever the gas held in the gaseous structures may be, it is proportionately warmer, and therefore proportionately lighter than air."

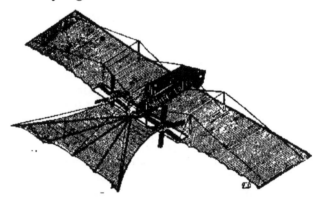

1842.—AËROPLANE INVENTED BY HEWSON, PROPELLED BY STEAM.

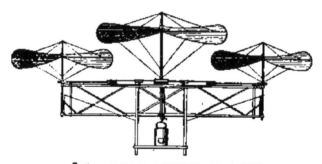

1845.—MACHINE INVENTED BY COSSUS.

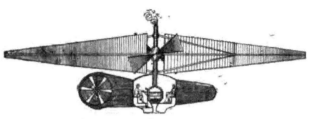

1874.

1878.—This year Professor Ritchell exhibited at Hartford, Connecticut, a flying machine, the result of many years' study.

The lifting power was obtained from a horizontal cylinder of gossamer cloth coated with india-rubber 25 ft. long by 13 ft. in diameter and weighing 66 lbs. A net-work of worsted bands enclosed the cylinder, which was connected to a strong brass tube 1½ ins. in diameter and 23 ft. long, to which the flying machine proper was attached. This consisted of an arrangement of hollow brass rods, very light and strong, which carried the gearing and a four-bladed fan or screw propeller, which could be rotated at the rate of 2,000 revolutions per minute; this propeller was 24 ins. in diameter and was worked by treadles from a small seat, and attached to this gearing was a vertical fan 22 ins. in diameter; the cylinder was filled with hydrogen gas, which made the machine floatable.

The Professor rose with this to a height of 250 feet, sailed over the Connecticut River, and worked his way back by means of the propeller.

Having exhausted the subject of flying machines, the next chapters are on Balloons, and the first persons who tested their theories by actual demonstration, and showed by the best of all proofs the possibility of men rising into the air, were the Messieurs Montgolfiers.

CHAPTER I.

1782 to 1784.

———+·+———

"There's something in a flying horse,
There's something in a huge balloon."
Wordsworth.

(*Translation.*)

AËROSTATIC MACHINES.

A SHORT EXTRACT from the Appendix to the
69th Volume of *The Monthly Review* by
M. Fanjas de St. Fond.

THE preface to the above volume contains a
short survey of what projects have formerly
been suggested for the purpose of floating heavy
bodies in the atmosphere; the principal which
are those of Lana, a Jesuit of Brescia; and of
Galieen, a Dominican of Avignon; both of which,
however, were, upon well-established principles,
found by theory to be impossible in the execution.

Due honour is paid to Mr. Cavallo, of
London, who in 1782, seemingly with a view to
this discovery, tried to fill bags of paper and

bladders with inflammable air: but failed in his attempts by the unexpected permeability of paper to inflammable air, and the too great proportional weight of the common sized bladders.

The honour of the discovery is certainly due to the brothers Stephen and Joseph Montgolfier, proprietors of a considerable paper manufactory at Annonay, a town about 36 miles south of Lyons, and their invention is the more to be admired, as it is not the effect of the discovery of a permanent elastic fluid lighter than the common air, but of properties of matter long known, and in the hands of many acute philosophers of this and of the last century.

They conceived that the effect they looked for might be obtained by confining vapours lighter than common air in an inverted bag, sufficiently compact to prevent their evaporation, and so light that when inflated its own weight, added to that of the enclosed vapour, might fall somewhat short of the weight of the air which its bulk displaces.

On these principles they prepared matters for an experiment: they formed a bag, or balloon, of linen cloth lined with paper, nearly spherical, about 35 ft. in diameter; its solid contents were about 22,000 cubic feet, a space nearly equal to that occupied by 1,980 lbs. of common air; of a mean temperature, on the level of the sea, the vapour was about half as light as common air, weighing 990 lbs. The balloon, together with a wooden frame suspended to the bottom which was to serve as ballast, weighed

490 lbs., whence it appears that the whole must have been about 500 lbs. lighter than an equal bulk of common air. This difference of specific gravity by which these bodies are made to rise is called *" their power of ascension."*

The 5th of June 1783 was fixed for this singular experiment, the States of Vivarais, who were then assembled at Annonay, were invited to the exhibition.

The flaccid bag was suspended on a pole 33 ft. high, and straw and chopped wool were burnt under the opening at the bottom: the vapour soon inflated the bag, and on a sudden this immense mass ascended into the air with such velocity that in less than ten minutes it appeared to be about 1,000 toises above the heads of the spectators, then the vapour escaping through some loopholes the great globe gradually descended with little damage to it.

II.—The Parisian philosophers resolved to use, instead of vapour, inflammable air. This being expensive, the author of the book now quoted set on foot a subscription and raised a sum sufficient to set to work, and they constructed a globe of taffetas glazed over with elastic gum, this was filled with inflammable air, produced from 1,000 lbs. of iron filings and 498 lbs. of vitriolic acid.

On the 29th of August 1783 the balloon was conveyed by night to the Champ de Mars Troops were drawn up to prevent disturbance, the concourse of people being immense; the cords were cut, and in less than two minutes the balloon

B

rose to a height of near 500 toises; it was then lost among the clouds; and in three-quarters of an hour fell in a field near Gonesse.

III.—M. Montgolfier, Junior, constructed a balloon of an elliptical form 70 ft. high and 40 ft. diameter, and on the 12th September 1783 it was filled with smoke, loaded to a weight of 500 lbs., and ascended fastened to ropes in the presence of the Deputies of the Royal Academy, but being a wet day it was not set loose.

Two days previous, on the 10th September, in the presence of the King, Queen, and Court, and all those who could get to Versailles, a balloon 57 ft. high and 41 ft. in diameter was sent up, the cage contained a sheep, a cock, and a duck: it remained in the air 20 minutes and fell in the wood of Vaucresson, the animals uninjured.

IV.—M. Montgolfier made a new balloon in a garden in the Faubourg St. Antoine 70 ft. high and 46 ins. diameter.

A gallery of wicker-work was around the aperture at the bottom, under which an open grate or brazier was suspended so that the fire might be fed to keep up the vapour.

On the 15th October M. Pilatre de Rozier, the most intrepid philosopher of the age, placed himself in the gallery, ascended 80 feet from the ground and there kept the balloon afloat by throwing straw and wool upon the fire.

On the 19th October he ascended a second time, about 250 feet, the balloon was then hauled down, and M. Giron de Villette placed himself in the gallery opposite M. de Rozier, and they were let up, and for some time hovered over Paris, in the sight of all its inhabitants at the height of 324 feet.

The following is part copy of a translation of a copy of a certificate dated at the Château de la Muettle near Paris :—

"21 November 1783.

"To-day, at the King's Palace, an experiment has been made of the aërostatique machine of M. Montgolfier.

"The Marquis D'Arlandes and M. Pilatre de Rozier were placed in the gallery; the machine rose in a majestic manner to a height of above 250 feet, and was soon out of sight; then it appeared at a height of 3,000 feet, crossed the Seine, and was visible all over Paris.

"They descended in an easy manner opposite the mill of Croulebarbe, without having experienced the least inconveniency, having still left in their gallery about two-thirds of their provisional stores; they might, therefore, if they had chosen, have gone over a space of treble the extent; their time was 25 minutes.

"The machine was 70 feet in height, 46 in diameter, and bore up a weight of 1,700 lbs. ·

"Signed by the Duc de Polignac, Duc de Guines, and others."

On the 1st December 1783 Messrs. Charles and Robert ascended in a globe 5 ft. 8 ins. in diameter.

M. Montgolfier cut the string; they carried with them a barometer, blankets and furs.

After nearly three-quarters of an hour they descended on the plains of Nesle, held a conversation with the Duc de Chartres, the Duke Fitz-James, Mr. Farrer, an English gentleman, and others. They then made a second ascent, and remained up 35 minutes, descending at about a league from the place where they set out.

FIRST LADY ASCENT.

On the 28th June 1784 an ascent was made at Lyon before the King of Sweden, then travelling as Count Haga.

Madame Thiblé accompanied the aëronaut, and she was the first lady who ever went up into the clouds.

On the 15th September 1784 an Italian named Lunardi made the *first ascent in Britain*, in a balloon from the Artillery Grounds, Moorfields, in the city of London.

It was intended that another gentleman should accompany him, but it was found that this would be too great a weight, so a smaller gallery was substituted; then Lunardi found that an accident had happened which would prevent the ascent; but the mob were hostile, and he, according to his own account, "almost deprived of his senses," and finding the injury trifling, was so alarmed that he forgot to take his instruments for observation with him.

"I threw myself into the gallery," said he, "determined to hazard no further accidents that might consign me and the balloon to the fury of the populace, which I saw on the point of bursting." The ropes were cut, and the balloon slowly and majestically arose.

After a voyage of a little over two hours, M. Lunardi landed in a field in Hertfordshire, among a crowd of rustics, who at first refused to give assistance to one who came on what they called "*the Devil's horse.*"

A country gentleman erected a stone on the spot where Lunardi alighted, and it bore this inscription :—

"Let posterity know, and knowing, be astonished, that on the 15th day of September 1784, Vincent Lunardi, of Lucca, in Tuscany, the first aërial traveller in Britain, mounting from the Artillery Ground in London, and traversing the regions of the air for two hours and fifteen minutes, on this spot revisited the earth.

"On this rude monument for ages be recorded that wonderous enterprise, successfully achieved by the powers of chemistry and the fortitude of man : that improvement in science which the Great Author of all knowledge, patronising by His Providence the inventions of mankind, hath graciously permitted, to their benefit and His own Eternal glory."

19th September 1784.

ROYALTY

made *its first ascent* in the person of the Duke of Chartres.

When 6,000 feet high the Duke began to be alarmed at a proximity to heaven which he had never calculated upon reaching, and abso-lutely *pierced the lower part of the silk bag with his sword* in order to get down the quicker.

This voyage was in the clouds—and thunder-clouds too—for five hours, and they travelled 135 miles.

CHAPTER II.

1785 to 1786.

——•——

"What divine monsters, O ye gods, are these
That float in air and fly upon the seas."

Dryden.

DOVER TO CALAIS.

January 7th, 1785.

The wind being N.N.W., Mr. Blanchard accompanied by Dr. Jefferies, took his departure for the Continent in his balloon, from the Castle at Dover, at 13 minutes past 1 from the large gun, known by the name of Queen Anne's pocket pistol. This being the sixth voyage performed by Mr. Blanchard in this balloon.

He took with him letters from the Prince of Wales, the Duchess of Devonshire and others to the principal personages of the French Court.

After an exciting time they entered France at 3 o'clock, descending in the Forest De Felmores, and arrived at Calais between 1 and 2 in the morning.

They expended all their ballast, the anchors, their coats and trousers, &c.

January 20th, 1785. Dublin.

About 40,000 persons collected at Ranelagh to see Mr. Crosbie ascend in his balloon, the cord was cut, he mounted awfully majestic, and in three minutes and a half an envious cloud secluded him from mortal sight, and all was solemn silence; in about twelve minutes he appeared descending at the northward, where he was carried in procession to the Earl Charlemont's.

The balloon and chariot were beautifully painted and the arms of Ireland emblazoned on them in superior elegance of taste.

Mr. Crosbie's figure was genteel, his aërial dress consisted of a robe of oiled silk lined with white fur, his waistcoat and breeches in one, of white satin quilted and morocco boots and a Montero cap of leopard skin.

The Duke of Leinster, Lord Charlemont, Right Hon. George Ogle, Counsellors Caldbeek, Downes and Whitestone, attended with white staves as regulators of the business of the day.

March 23rd, 1785.

Count Zambeccari and Sir Edward Vernon, ascended in a balloon from the vicinity of the Tottenham Court Road, it snowing at the time.

On the eve of their departure a Miss Grice of Holborn offered to accompany them; it was accepted and she entered the car, but it was found that the balloon would not rise, and she had to give up the pleasure of an ascension.

On her quitting the car, the balloon immediately ascended with amazing velocity and
was out of sight in a few minutes.

The Count and Sir Edward descended safe
in King's Fields, near Horsham in Sussex,
about 5 o'clock the same day, being 37 miles in
little more than an hour.

Admiral Vernon was probably the *first Admiral*
to navigate the air.

April 19th, 1785.

Young Decker, an intrepid youth of scarcely
seventeen years, ascended in a balloon from
Bristol, and landed 3 miles from Chippenham,
a circuit of about 30 miles, which he performed
in an hour and seven minutes.

June 14th, 1785.

Mr. Sadler proposed to ascend in a balloon
from Oxford, Colonel Fitzpatrick with him : but
as the balloon would not mount Mr. Sadler got
out and the Colonel ascended alone.

In his passage the Colonel had not expended
any of his ballast; but descended in consequence
of a rent near the bottom of the balloon, occasioned by the expansion of the internal air,
which was not discovered by the Colonel till
after he had reached the ground.

He descended near Kingston Lisle, opposite
the White Horse Hills, Berkshire, without the
least injury.

June 15th, 1785.

A balloon was launched from Boulogne in
France, which took up Mons. Pilatre du Rosière

and M. Romain, and when they were at a great height the balloon took fire, burnt the cords by which the car was suspended, and both were dashed to pieces in a manner too shocking to mention.

July 25th, 1785.

Extract from a letter of this date from Major Money of Norwich :—

"On Saturday last I ascended about 5 o'clock in the afternoon from this place with a balloon and was driven out to sea, not being able to let myself down, from the valve being too small. After blowing about for two hours I dropped into the sea; the balloon was torn and like an umbrella over my head. A Dutch vessel, taking me for a sea monster, left me to my fate."

"I exerted myself to preserve life by keeping the balloon floating over my head, sinking inch by inch as it lost its power to keep me out of the water.

"I was breast high when taken up by a revenue cutter at half-past eleven at night, and so weak I was obliged to be lifted out of the car into the ship. I was put to bed, fell asleep, and did not wake till six the next morning.

"We landed at Lowestoft at eight. Any man with less strength than myself must have perished."

April 11th, 1786.

Blanchard performed his 27th aërial excursion. He took his departure from Douay in Flanders, and descended near l'Etoile, a village in Picardy, a voyage of 90 miles in as many minutes.

June 18th, 1786.

This day took place the ascension of the physicist, Jester. After starting from Paris, alone in a balloon of small dimensions, filled with hydrogen, he came down in a village of Montmorency. He descended in a field of nearly ripe corn, and the proprietor, indignant at the damage done, came out with a number of his peasants to clamour for compensation.

Jester refused to pay anything on the ground that the harm done was accidental, whereupon the labourers, with the view of dragging him before the local magistrate, seized hold of one of the ropes and towed the balloon after them, whilst a farm-boy, in order to prevent Jester from escaping, climbed into the car.

After going some distance Jester began to think that he should be forced to pay, so he threw out a large portion of his ballast, opened his knife and cut the rope by which he was being hauled before the magistrate, upon which, to the stupefaction of the rustics, the balloon rose swiftly into the air and disappeared in the clouds.

It is said that when the farm-boy descended some time later with the aëronaut his hair had turned grey.

September 20th, 1786.

On this date Lunardi's ascent from the Spital - ground, Newcastle-upon-Tyne, was pro-ductive of a melancholy accident.

The balloon was about one-third full, when Lunardi went to pour into the cistern the rest of the oil of vitriol destined for the purpose.

This having caused a strong effervescence, generated inflammable air with such rapidity, that some of it escaped from two different parts of the lower end of the apparatus and spread among the feet of several gentlemen who were holding the balloon, and who were so alarmed that, leaving it at liberty, they ran from the spot.

The balloon rose with great velocity, carrying up with it Mr. Ralph Heron, a gentleman of the town.

This unhappy victim held a strong rope which was fastened to the crown of the balloon twisted about his hand, and could not disengage himself when the other gentlemen fled. He was elevated about the height of St. Paul's cupola, when the balloon turned downward, and the unhappy gentleman fell to the ground.

He fell upon very soft ground, but died about an hour-and-a-half after the fall.

CHAPTER III.

1802 to 1811.

——•——

June 28th, 1802.

Mr. Garnerin with Captain Snowden, R.N., ascended from Ranelagh in his balloon, which for neatness of construction as well as for its admirable philosophic principles far surpassed anything of the kind ever before witnessed in this country.

The descent was made at six in the evening four miles from Colchester, thus voyaging sixty miles under an hour. In order to descend there was a valve which opened inwardly and which was operated by pulling a cord, and the balloon sank in proportion to the quantity of gas let out. The diameter of this balloon was 20 feet.

DAMAGES CLAIMED FOR DESCENT.

August 3rd, 1802.

Mr. Garnerin ascended from Vauxhall Gardens, his wife and a gentleman were with him; the first time for fifteen years since a lady had ventured, in this country, to soar the empyrean height.

Ascending to a considerable height, Garnerin let fall from the car a small parachute to which was suspended a cat. At a quarter-past eight the aëronauts descended in Lord Rosslyn's paddock on the top of Hampstead Hill. The descent of the cat with its little vehicle was gradual and perfectly safe. It fell into the garden of a Mr. C—— of Hampstead, who insisted on receiving three guineas for indemnification of the trespass committed in his grounds by poor puss—and the parachute.

September 21st, 1802.

An extraordinary display of aëronautical dexterity was this day prepared with consummate skill, and executed with admirable intrepidity. The balloon was inflated on the parade ground of the St. George's Volunteers, near Grosvenor Square.

At ten minutes to five Mrs. R. B. Sheridan launched a pilot balloon, which in seven minutes was completely out of sight. Preparations were then made for launching the larger balloons to which a parachute was attached by cordage to the central tube about four feet above the basket, thus the only connection between the balloon and the parachute was formed by the rope passing through the central tube, which being cut from below, the latter was left to its proper action.

Garnerin ascended and was visible from every house in the Metropolis which had a

northern aspect. He evidently wished to pro-
long his stay for the gratification of the people
by opening the ·valve of the balloon, and on
each discharge of the inflammable air the
balloon, illuminated by the setting sun, appeared
to be surrounded by a nimbus or glory, such
as is seen to surround the heads of saints, &c.,
in paintings of scriptural subjects.

Garnerin, when at a height of 4,000 feet,
cut the rope, and the parachute was seen to
descend with the utmost velocity: he landed in
safety in a field near St. Pancras Church. He
received only a slight hurt on one side of his
face, from being thrown out of the basket, for
though this had a false bottom, so constructed
as to break the fall, it had little effect on the
velocity of his lateral descent.

December 4th, 1802.

Citizen Oliveri, a physician from Paris,
ascended from Orleans in a Montgolfier balloon,
and unhappily fell a victim to his imprudence.
He disappeared in the clouds in less than three
minutes, and his body was soon after found about
three miles from the town.

The balloon took fire in the air and the in-
discreet aëronaut, of course, fell precipitately to
the earth.

A BALLOON DUEL.

1808.—Perhaps the most remarkable duel ever
fought took place in 1808. M. de Grandpré with
M. le Pique, had a quarrel arising out of jealousy

concerning a lady engaged at the Imperial Opera.

They agreed to fight a duel to settle their respective claims, and in order that the heat of angry passion should not interfere with the polished elegance of the proceedings they postponed the duel for a month, the lady agreeing to bestow her smiles on the survivor of the two, if the other was killed.

The duellists were to fight in the air, and two balloons were constructed exactly alike. On the day fixed De Grandpré and his second entered the car of one balloon, Le Pique and his second that of the other, it was in the garden of the Tuileries, amid an immense concourse of spectators. The gentlemen were to fire not at each other, but at each other's balloons in order to bring them down by the escape of gas, and each took a blunderbus in his car.

The ropes were cut, the balloons ascended, and about half a mile above the surface of the earth, a preconcerted signal for firing was given. M. le Pique fired and missed. M. de Grandpré fired and sent a ball through Le Pique's balloon, which collapsed and the car descended with frightful rapidity, and Le Pique and his second were dashed to pieces.

De Grandpré continued his ascent and landed safely at a distance of seven leagues from Paris.

July 30th, 1811.

Madame Blanchard in one of her ascents from Paris in a balloon was caught in a storm of hail and rain, but, notwithstanding, ascended so high that she was lost in clouds and whirlwinds, and did not alight from her balloon, near Vincennes, till between 6 and 7 in the morning, the day after she rose from Paris.

In consequence of the prodigious height to which the balloon ascended, Madame Blanchard fainted, and continued insensible for some time. Her ascension occupied fourteen hours and a half.

October 1st, 1811.

M. Girard ascended from Florence in a balloon. In half-an-hour he lost sight of the earth, and found himself at an elevation of 15,000 feet.

The balloon still continued to rise, when M. Girard finding his limbs benumbed by the extreme cold, and himself nearly overpowered by sleep, manœuvred to descend; but, perceiving beneath him the Mediterranean Sea, he rose again, and suffered still more from the excessive cold. He journeyed thus in the heavens until 2 o'clock in the morning: he then perceived land, and descended safely at St. Gascians, having from the moment of his ascension been absent nine hours.

CHAPTER IV.

1811 to 1817.

October 7th, 1811.

Mr. Sadler ascended in his balloon with a companion at Birmingham, and in an hour and twenty minutes they were wafted the space of more than 100 miles, to Heckington, in Lincolnshire, where the balloon, in its descent, catching in a tree, was torn to pieces, but the voyagers escaped without injury.

June 29th, 1812.

Mr. Sadler made his twenty-third ascent at Manchester, and alighted at Oakwood, about six miles from Sheffield. He made the passage in 48 minutes, so that he must have travelled at the amazing rate of a mile in a minute.

A PERILOUS VOYAGE.

October 1st, 1812.

Mr. Sadler ascended from Belvidere House, in Dublin, with the wind at S.W., and in

35 minutes had sighted the mountains of Wales, at three o'clock he was nearly over the Isle of Man, and at four he was in view of the Skerry Lighthouse. The wind shifted, and, seeing five vessels beating down Channel, and in hopes of their assistance, he descended and precipitated himself into the sea. The vessels took no notice of him. He threw out a quantity of ballast, and quickly regained his place in the air.

Long after he observed a vessel which gave him to understand by signal that they would help him, but could not reach him. He again descended and tied his clothes to the grappling-iron and sunk them to keep him steady; still the balloon was carried away so fast that he expelled the gas; upon that escaping the car actually sunk and he had then nothing but the netting to cling to. This perilous situation, and the fear of getting entangled, deterred the men from coming near him until, being in danger of drowning, Mr. Sadler begged they would run their bowsprit through the balloon and expel the remaining gas. Having done this they threw out a line, which he wound round his arm. He was dragged a considerable way before they could get him on board, quite exhausted.

The ship was the "Victory," a herring-fisher from Douglas.

April 17th, 1813.

Mr. Cameron ascended from Glasgow. The balloon went up in fine style and descended at Falnash, in the county of Roxburgh, having travelled 74 miles in 1 hour and 20 minutes.

CURIOUS ASCENT.

September 7th, 1813.

Cheltenham. Mr. Sadler proposed to ascend from this place, but the balloon had not power to rise with him. The balloon was made of white and crimson silk, in the shape of a Windsor pear, but not upon such a large scale as was intended.

After the car had been properly fastened, William, aged 16, the son of Mr. Sadler, entered the car, the ropes were loosened, and the balloon rose in the most magnificent style. Mr. Sadler set off in a carriage to follow the balloon, and his eldest son on horseback.

When at the highest elevation the atmosphere was oppressive, and a thick fall of snow beat against the balloon with so much violence that it was with the utmost difficulty that Sadler could open the valve, when he descended over Burford. He rose again to avoid descending in Wedgewood Forest, and at ten minutes before six descended in a field a short distance from Chipping Norton.

The first man that approached him was armed with a pitchfork, who cried, " Lord, Sir, where do you come from ? "

AN EXCITING ASCENT.

August 1st, 1814, was the day fixed for the grand National Jubilee, being the centenary of the accession of the illustrious family of Brunswick to the throne of this Kingdom, and the anniversary of the battle of the Nile.

Mr. Sadler, junior, ascending in a balloon from the Green Park. When the cords which held the balloon were ready to be cut, it was found that the fastening which secured the network to the valve at the top of the balloon had by some means been disengaged and was held only by a single twine. However, he ascended about twenty-four minutes past six.

On passing over Deptford at a considerable height, Mr. Sadler went through a cloud which left behind it, on the railing of the car and on various parts of the balloon, a thick moisture which soon became frozen, and Mr. Sadler for a short time felt the cold as intense as in winter.

Immediately over Woolwich the string which fastened the net suddenly broke, and the main body of the balloon was forced quickly through the aperture, nearly 18 feet. Mr. Sadler caught the pipe at the bottom of the balloon, and by hanging on it and the valve line he prevented the balloon from further escaping. The valve, which had resisted every attempt to open it in consequence of being frozen, at this time gave way, and suffered the gas to escape.

The balloon was apparently falling into the Thames at Sea Reach, when a sudden shift of wind carried it over the marshes on the Essex side. Mr. Sadler seized the opportunity of making a gash in the balloon with his knife, which the wind widened and occasioned the escape of the gas in great quantities, and his descent on this account was rather more precipitate and violent than he could have wished.

Mr. Sadler landed at Mucking Marshes, 16 miles below Gravesend, without sustaining any other injury than a slight sprain.

September 20th, 1815.

Madame Garnerin ascended in a balloon from the gardens of the Tivoli. A number of persons of distinction were present, including the King of Prussia and the Prince Royal, his son.

Madame Garnerin wore a wreath of flowers on her head, and was clothed in a simple white robe. At six o'clock she threw herself into the car, and rose amid the acclamations of the people. The signal to cut the cords of the parachute was given by M. Garnerin, her father, by means of a *boite,* which exploded two minutes after her departure; but she was too elevated to hear the report, as she was not detached from the balloon till four minutes and a half after her departure. The elevation was so considerable that the descent occupied more than five minutes.

BALLOON ASCENT IN DAYS GONE BY.

89 Years ago.

The following account of an aërial voyage in the month of September 1817 is given by the voyager himself, Prince Pückler Muskau, in a work entitled *Tutti Frutti,* published in 1834. He writes:—

"I had scarcely recovered from a severe illness when M. Reichhard, the aëronaut, came to Berlin, and paid me a visit for the purpose of receiving introductory letters.

"His interesting narrative awakened in me an irresistible desire to soar once in my life to the empire of the eagle. He interposed no obstacle to the gratification of my wishes, and we decided that he should construct a balloon at my expense—the different items amounted to 600 rix dollars (£40). But even at this rate the pleasure I enjoyed was cheaply purchased.

"The day which we selected was one of the most heavenly that could be imagined : scarcely a cloud was to be seen in the firmament; half the population of Berlin were assembled in the streets, squares, and on the roofs of the houses.

"We entered the car, and out of the centre of this motley multitude, ascended majestically towards the heavens. Our frail aërial bark, not much larger than a child's cradle, was surrounded by a network as a precaution against any giddiness that might ensue; but notwithstanding the weakness which remained after my indisposition, I did not experience the slightest disagreeable sensation.

"As we gently and slowly ascended, I had sufficient time to salute and receive in turn the farewell salutations of my friends below. No imagination can paint anything more beautiful than the magnificent scene now disclosed to our

enraptured senses.   The multitude of human
beings, the houses, the squares and streets, the
highest towers gradually diminishing, while the
deafening tumult became a gentle murmur, and
finally melted into a death-like silence.

"The earth which we had recently left lay
extended in miniature relief beneath us, the
majestic linden trees appeared like green fur-
rows, the river Spree like a silver thread, and
the gigantic poplars of the Potsdam Allée, which
is several leagues in length, threw their shadow
over the immense plain.   We had probably
ascended by this time some thousand feet, and
lay softly floating in the air, when a new and
more superb spectacle burst upon our delighted
view.

"As far as the eye could compass the horizon
masses of threatening clouds were chasing each
other to the immeasurable heights above, and,
unlike the level appearance which they wear
when seen from the earth, their entire altitude
was visible in profile, expanded into the most
monstrous dimensions—chains of snow-white
mountains wrought into fantastic forms, seemed
as if they were tumbling headlong upon us.
One colossal mass pressed upon another, encom-
passing us on every side, till we began to ascend
more rapidly, and soared high above them, where
they now lay beneath us, rolling over each other
like the billows of the sea when agitated by the
violence of the storm, obscuring the earth entirely from our view.   At intervals the fathomless
abyss was occasionally illumined by the beams

of the sun, and resembled for a moment the burning crater of a volcano; then new volumes rushed forward and closed up the chasm; all was strife and tumult. Here we beheld them piled on each other white as the drifted snow, there in fearful heaps of a dark watery black, at one instant rearing towers upon towers, in the next creating a gulf, at the sight of which the brain became giddy, dashing eternally on-ward, in wild confusion.

"I never before witnessed anything compar-able to this scene even from the summit of the highest mountains; besides, from them the continuing chain is generally a great obstruction to the view, which, after all, is only partial; but here there was nothing to prevent the eye from ranging over the boundless expanse.

"The feeling of absolute solitude is rarely experienced upon earth, but in these regions, separated from all human associations, the soul might almost fancy it had passed the confines of the grave. Nature was noiseless, even the wind was silent, therefore receiving no opposi-tion, we gently floated along, and the lonely stillness was only interrupted by the progress of the car and its colossal ball, which, self-propelled, seemed like the rock-bird fluttering in the blue-ether.

"Enraptured with the novel scene I stood up, in order to enjoy more completely the superb prospect, when M. Reichhard with great *sang-froid* told me I must be seated, for that, owing to the great haste with which it had been

constructed, the car *was merely glued,* and therefore might easily come asunder unless we were
careful. It may readily be supposed, that, after
receiving this intimation, I remained perfectly
quiet.

"We now commenced descending, and were
several times obliged to throw out some of the
ballast in order to rise again. In the meantime we dipped insensibly into the sea of
clouds which enveloped us like a thick veil,
and through which the sun appeared like the
moon in Ossian. This illumination produced a
singular effect, and continued for some time
till the clouds separated, and we remained
swimming about beneath the once more clear
azure heavens.

"Shortly after we beheld, to our great
astonishment, a species of ' *fata morgana* '
seated upon an immense mountain of clouds,
the colossal picture of the balloon and ourselves
surrounded by myriads of variegated rainbow
tints. A full half hour the spectral-reflected
picture hovered constantly by our side. Each
slender thread of the net-work appeared distended to the size of a ship's cable, and we
ourselves like two tremendous giants enthroned
on the clouds.

"Towards evening it again became hazy,
our ballast was exhausted, and we fell with
alarming rapidity, which my companion ascertained by his barometer, although it was not
apparent to the senses.

"We were now surrounded for some time by a thick fog, and as we rapidly sank through it we beheld in a few minutes the earth beneath glowing in the most brilliant sunshine and the towers of Potsdam, which we distinctly beheld, saluted us with a joyful carillon. Our situation, however, was not so full of festivity as our reception. We had already thrown out our mantles, a roasted pheasant, and a couple of bottles of champagne, which we had taken with us for the purpose of supping in the clouds, laughing heartily at the idea of the consternation which this proceeding would cause in any of the inhabitants of the earth, who happened to be sleeping upon the turf, in case the pheasant should fall into his mouth, and the wine at his feet, but we could not forbear hoping that it would not descend upon his head, as, instead of an agreeable excitement to his brain, it would act the part of a destroying thunderbolt.

"We were ourselves, like the other articles, tumbling, but to our great consternation we saw nothing beneath us but water, the various arms and lakes of the river Havel, only here and there intermixed with wood, to which we directed our course as much as possible. We approached the latter with great velocity, which appeared to me from the height like an insignificant thicket. In a few seconds we were actually hanging on one of the branches of the shrubs, for such I really believed them to be, in consequence of which I commenced making the necessary arrangements to descend, when Reichhard, with great animation, called out, 'In God's name, stir not, we are entangled on the tops of an immense pine!' I could hardly believe my eyes, and it required the lapse of several seconds to convince me that what he asserted was really true, having entirely lost in a few hours the capacity of measuring distance.

"We were most certainly perched on the highest branches of an enormous tree, and the means to descend set our inventive powers at defiance, we called or rather shouted for help, first in solo, then in duetto, till we began to fear that we should be obliged to support our character of birds by roosting in the tree, for night was fast approaching.

"At length we saw an officer riding along the high road, which caused us to renew our cries with redoubled vigour; he paused, but thinking it might be robbers, who were endeavouring to inveigle him into the wood, galloped off with the rapidity of lightning, but as we continued vociferating, he gave a heaven-directed glance, discovered us, raised himself in the saddle, reined in his horse, and with outstretched neck and distended eyes, endeavoured to ascertain, if possible, the nature of the singular nest he beheld in the gigantic pine. At length, having satisfied himself that we were really not of the winged creation, he procured men, ladders, and a carriage from the neighbouring town. But as all this consumed no inconsiderable space of time, we remained perched in mid air, and it was quite dark when we arrived at Potsdam with our balloon, which, by the way, was very little injured. We took up our abode at the Hermit Hotel, at that time badly conducted, where we alas! had ample reason to regret the loss of our supper."

CHAPTER V.

1819 to 1836.

———◆———

1825.—Madame Blanchard ascended in a luminous balloon, ornamented with artificial fireworks, in the evening from the Tivoli Gardens; she was dressed in white, wearing also a white hat with feathers.

The signal being given, the balloon rose gently; but by throwing out ballast Madame caused it to ascend more rapidly. The Bengal fire-pots illuminated this brilliant ascent.

Suddenly the balloon entered a light cloud which completely extinguished the fire-pots. Madame Blanchard then ignited the artificial fireworks which produced the effect expected, when some of the flying fuses were seen to direct themselves perpendicularly towards the balloon and the fire communicated with its base. A frightful brilliancy instantly struck terror into all the spectators, leaving no doubt of the deplorable fate of the aëronaut.

The lifeless body of Madame fell from a height of more than 400 feet in the Rue de Provence; the body was still in the car, being caught in the cords by which it was attached to the balloon.

AËRIAL VOYAGE TO NASSAU.

November 7th, 1836.

Numerous remarkable voyages in the regions of air have been made since the art of aërostation was originally established in France; but the most extraordinary voyage was that performed in November, 1836, by Messrs. Charles Green, Monck Mason, and Robert Holland. The chief object of this expedition was to test the value of certain improvements which Mr. Green's long practice in the art had led him to suggest.

Mr. Monck Mason describes the balloon as resembling a pear; in height over 60 feet, breadth about 50 feet, capable of containing more than 85,000 cubic feet of gas and competent to raise 4,000 lbs.

The balloon left Vauxhall Gardens at half-past 1 p.m.; provisions for a fortnight's consumption were taken; a ton of sand as ballast, barometers, and passports to various parts of the Continent, whither they intended to direct their course.

On arriving above Canterbury a small parachute was lowered with a letter addressed to the mayor, which was delivered to him; they passed directly over Dover and entered upon their course over the sea. They crossed the Straits in about an hour and at an altitude of nearly 3,000 feet, when they had supper. They had some lime, which they slaked with water, which gave sufficient heat to afford them a dish of warm coffee.

They passed over the city of Liége, and effected a safe landing six miles from the town of Weilberg, in the Duchy of Nassau, distant from London upwards of 500 miles.

The balloon was brought back to Vauxhall Gardens, having been formally christened the " Nassau."—*From Mr. Monck Mason's Book.*

CHAPTER VI.

1850 to 1858.

—••—

A PERILOUS ASCENT.

June 29th, 1850.

M. Barral, a chemist, and M. Bixio, a member of the Legislative Assembly, ascended from the garden behind the observatory at Paris. The balloon was old and in bad condition, but within two minutes from the time it was liberated it was plunged in the clouds and out of sight, 15 minutes after it was 14,200 feet above sea level. Rain wetted the balloon, and saturated the cordage, a few minutes after the height was 19,700 feet. The moisture had frozen upon the thermometer, and while M. Barral was in the act of wiping the icicles from it, he turned his eyes upwards and beheld what would have made the stoutest heart quail with fear.

To explain, they were nearly 20,000 feet above the surface of the earth, and a mile above the highest strata of clouds.

As it was intended to ascend to an unusual height, the balloon at starting was not nearly

D

filled with gas, and yet very nearly filled the network which enclosed it.

It seems strange that the scientific men present did not foresee that when it would ascend into a highly rarefied atmosphere it would necessarily distend itself to such a magnitude that the netting would be utterly insufficient to contain it. Such effect now disclosed itself to the astonished and terrified eyes of M. Barral. The balloon had so swelled as to completely fill the netting, forced itself through the hoop under it, from which the car was suspended; in short, the inflated silk nearly touched the heads of the voyagers. The valve was placed in a sort of sleeve; on looking for this, it had disappeared, it had been gathered up in the network above the loop: to reach it, it would have been necessary to force a passage between the inflated silk and the hoop. The access to, and the play of the valve, had been fatally overlooked at starting, and one thing of two must happen, either that the inflated silk would descend and suffocate them, or the balloon would burst. M. Barral climbed up the side of the car and the network suspending it, and forced his way through the hoop, so as to catch hold of the valve sleeve: in doing this he produced a rent in the silk below the loop and over the car. The hydrogen gas issued with terrible force from the balloon, and the voyagers found themselves involved in an atmosphere of it, and they were nearly suffocated: the barometer showed that they were falling with frightful rapidity.

M. Barral proceeded to examine the balloon and found a rupture had taken place 5 feet in length along the equator of the machine, through which the gas was escaping in immense quantities. To check the descent required coolness and skill, they retained nine or ten sand bags, but cast out blankets, clothing, shoes, wine, and all except philosophical instruments. On arriving within a few hundred feet of the earth the nine sand bags were dropped, and this probably saved their lives. The balloon reached the ground, the car struck among some vineyards near Lagny; the voyagers were afraid to leap from the car. M. Barral threw his body half down from the car, laying hold of the vine stakes as he was dragged along, directing M. Bixio to hold fast to his feet : in this way their united bodies formed a sort of anchor ; the arms of M. Barral playing the part of the fluke, and the body of M. Bixio that of the cable. The labourers in the vineyard pursued and finally succeeded in capturing the balloon.

The entire descent from the altitude of 20,000 feet was effected in seven minutes, being at the rate of 50 feet per second.—"*The Gifts of Science to Art,*" *Dublin University Magazine.*

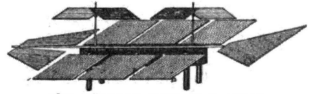

1851.—AËROPLANE MODEL BY M. AUBAUD.

A RASH LADY.

June, 1851.

The *Revue des Deux Mondes* records a trip taken
by M. Godard, in the "Eagle" balloon, his companion, Madame the Comtesse de S., Count Alexis
de Pomereu, and a friend.

The air was calm and the sky pure, the party
in high spirits and without the least thought of
danger. "Not one of us," said M. Matzneff, "felt
any acceleration in the beating of his heart," and
for a long time they enjoyed the panoramic view
of the great city beneath, which inspired the
sentiment. "Viewing human things from such a
height, one feels that life is more insignificant and
nature greater, the instinct of preservation recalling to the earth, but still more powerful the
attraction toward the sky."

These contemplations were interrupted by the
lady, who, in sportive humour, amused herself by
causing the car to "oscillate capriciously" with
sudden shocks, and "at times leaning over the
edge, defying the abyss, and seriously compromis-
ing our equilibrium. At last, yielding to the
respectful injunctions of the party, she consented
to relinquish her experiments."

After this they dined "as comfortably as in one
of the saloons of the Frères-Provençaux," and
drank healths and talked of the possibility of
directing balloons until it was time to descend.

As they approached the earth, the guide-rope,
150 metres long, was lowered, and "seized by some
labourers, who drew us without a shock to the

middle of their field, near the village of Bussy-le-Long, distant about sixty-six miles from Paris, the journey having occupied three hours and a half."

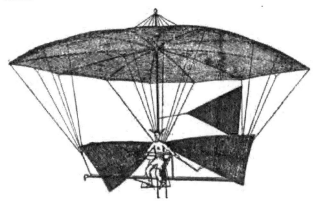

1852.—PARACHUTE, GOVERNED BY WINGS, IN WHICH
THE INVENTOR, LETOUR, WAS KILLED.

DAMAGE CLAIMED FOR DESCENT.

In *My Life and Balloon Experiences*, by Mr. Henry Coxwell, he relates "that in 1854 he made an ascent with the late Mr. B. O. Conquest and Mr. John Allan, from London, and as they did not wish to go far from home the descent was made at Barnet.

"A farmer was asked to take a glass of Mr. Conquest's champagne. 'No,' said he, 'hang your wine; I want your money.' 'What for?' 'What for!' he replied, 'why, damages!'

'Damages! where are they to be found?' Said
the farmer: 'Do you think that a lot of people
can come into my field without doing mischief?
I want to be paid.' I whispered Mr. Conquest
that he had better get out and work his way
home as best he could, so I let off a quantity
of gas, and asked two or three men to hold on,
and then asked the farmer to prove damages
and not imagine it. 'That can be done presently,' he replied. 'I want £3, and I shall

1857.—LE BRIS. THESE WINGS WERE LOWERED
BY MEANS OF LEVERS.

stick to you and your balloon until it is paid.'
Mr. Conquest cried out, 'Hands off there! I'll
settle this matter.' In a trice up shot the balloon, leaving the farmer amazed. He at once
pounced on Mr. Conquest, who replied: 'I am

only a passenger, and merely require twelve good men and true, at a shilling apiece, to show me the way to the railway station.'

"'Here you are, sir, to any number; and as to that greedy hound, if he lays hands on you or interferes in any way, we'll duck him in our horse-pond. This way, sir, for the up train, no one shall harm you.'

"So the old selfish farmer was done."

------

fiction.

THE BABES IN THE CLOUDS.

COMETS AND BALLOONS.

(AN AMERICAN STORY.)

"The year 1858 was the Great Comet year.

"One pleasant Saturday afternoon during the comet's appearance, an aëronaut, after a prosperous voyage, descended upon a farm near a large market town in one of the Western States. He was soon surrounded by a group of the farmer's family and labourers. The rope with anchor was in the hand of the aëronaut, its car but a foot or two above the ground. The farmer led the balloon to his house and said he could 'hitch it' to his fence, but before he thus secured it his three children, aged 10, 8, and 3, begged him to lift them 'into that big basket' that they might sit on 'those pretty red cushions,'

" The aëronaut's attention was diverted by curious questioners from a near farm, and the rash father lifted his darlings one by one into the car. Chubby little Johnny proved 'the ounce too much' for the machine and brought it to the ground, and then, unluckily, not the baby, but the eldest hope of the family was lifted out. The relief was too great for the monster; it rose, jerked the rope out of the farmer's hand, and with a wild bound mounted into the air! Two little white faces peeping over the edge of the car, whose piteous cries of 'Papa!' 'Mamma!' grew fainter and fainter up in the air. The father and mother were frantic with grief; the aëronaut strove to console them, saying: 'The balloon would come down.'

"Jennie, the eldest, took off her apron and wrapped it about the child, saying tenderly: 'This is all sister has to make you warm, darling, but she'll hug you close in her arms, and we will say our prayers and you shall go to sleep.' So the two baby wanderers, alone in the wide heavens, unawed by the presence of the great comet and the millions of unpitying stars, sobbed out 'Our Father,' and that quaint prayer—

Now I lay me down to sleep,
I pray the Lord my soul to keep;
If I should die before I wake
I pray the Lord my soul to take.

" 'There, God heard that easy, for we are close to him up here,' said little Johnnie. Soon they were in perfect peace, sleeping soundly. Poor babes in the clouds!

"At length a happy chance—or say, Providence—guided the little girl's hand to a cord connected with the valve, something told her to pull it; at length the balloon began to sink slowly and gently as though let down by tender hands.

"The sun had not yet risen, but twilight had come, and the little girl saw the dear old earth coming nearer, 'rising towards them,' she said, and presently the car caught fast in the topmost branches of a tree, yet they saw no house near where help might come.

"Farmer Burton, who lived on the edge of his own private prairie, was a famous sleeper in general, but on this morning he turned and turned and could sleep no more, so he said to his wife, 'It's no use, I'll get up, and have a look at the comet.' Then she heard a frightened summons to the door—a strange shape was seen hanging in a large pear tree—a trembling little voice, 'Please take us down, we are very cold.' A second voice, 'And hungry too, please take us down, we are Mr. Harwood's little boy and girl and we are lost in a balloon.'

"The farmer got hold of the dangling rope and pulled down the balloon, lifted the children out, and sent a mounted messenger to their home, and the parents arrived with a covered hay waggon.

"Joy bells were rung in the town, and in the farmer's brown house the happiest family on the continent thanked God that night."

It would seem that this comet had some occult maddening influence on balloons, for in another Western State an involuntary ascension similar to the last was made, but more tragical in its termination.

" An aëronaut while repairing the net-work of his balloon was seated on a slight wooden cross-piece, suspended under it, the car having been removed, and the balloon being held in its position a few feet from the ground by merely a rope in the hand of an assistant; from a too careless grasp · this rope escaped, and in an instant the gigantic bubble shot upward carrying the aëronaut on his frail support. A rider more helpless than Mazeppa bound to his ukraine steed : a voyager more hopeless than a shipwrecked sailor afloat on a spar in mid-ocean.

" The aëronaut was known to be of uncommon nerve and presence of mind, and it was hoped that he might manage to operate on the valve, or puncture a small hole in the balloon and thus effect a descent.

" We waited in vain—we gave him up—only wifely love hoped on, and looked and waited. At last the wreck of the balloon was found—that was all. Later some children nutting found a strange dark mass that looked like a heap of old clothes, but that there was a something shapeless and fearful, holding it together.

" It was thought that the aëronaut parted company from the balloon by loosening his hold on the cord above him in desperate efforts to open the valve, or had become unnerved by the awful silence of the upper night and by the comet's fearful companionship and wearily let go his hold, to drop earthward."

The above are taken very briefly from an article in *All the Year Round,* 1868.

CHAPTER VII.

1862 to 1870.

August 21st, 1862.

Mr. Glaisher this day ascended from the Crystal Palace in his balloon, and thus describes cloudland :—

"At 4.57 we were in cloud, surrounded on every side by white mist; gradually we emerged from the dense cloud into a basin surrounded with immense mountains of cloud rising far above us, and shortly afterwards we were looking into deep ravines, bounded with beautiful curved lines, the sky immediately overhead was blue, dotted with cirrus clouds: as we ascended, the tops of the mountain-like clouds became silvery and golden.

"At 5.1 we were level with them, and the sun appeared flooding with golden light all the space we could see for many degrees both right and left, tinting with orange and silver all the remaining space around us. It was a glorious sight indeed.

" Here arose shining masses of silvery heaps, there large masses of cloud in mountain chains, rising perpendicularly from the plain, dark on one side and silvery and bright on the other, with summits of dazzling whiteness : each large mass of cloud cast behind it its shadow, and this circumstance, added to the very many tints, formed a scene at once most beautiful and sublime." ✕

1862.—Mr. Glaisher and Mr. Coxwell made an ascent, and rose to the unprecedented height of fully seven miles, but both nearly lost their lives. At 29,000 feet from the earth Mr. Glaisher found his arm suddenly become powerless followed by a loss of all muscular power and inability to speak, and he was for some time insensible.

Mr. Coxwell found it necessary to clamber into the rings to disentangle the valve-rope. It was piercingly cold, with hoar frost round the neck of the balloon ; while on attempting to return his hands were frozen. Placing his arms on the ring he dropped down and found Mr. Glaisher laying insensible against the side of the car; he attempted to help him but could not, and, feeling insensibility coming over him too, "he became anxious to open the valve, but having lost the use of his hands, he could not do this, but succeeded by seizing the cord with his teeth, and dipping his head two or three times until the balloon took a decided turn downwards." They made a very rapid descent, recovered themselves, and landed safely. —*Travels in the Air*, J. Glaisher, F.R.S., 1871.

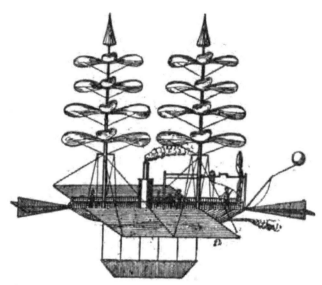

1863.—THIS INVENTION, ATTRIBUTED TO M. DE LA LAUDELLE, CONSISTED OF A COMBINATION OF INCLINED PLANES, AND PARACHUTE SYSTEM PROPELLED BY STEAM.

THE FIRST BALLOONIST SOCIETY.

January 12th, 1866.

This day a meeting was held at the Duke of Argyll's residence at Campden Hill, to found an *Aeronautical Society of Great Britain.*

The Duke was chosen chairman, the Duke of Sutherland vice-chairman, Lord R. Grosvenor another vice-chairman, and Mr. Glaisher, treasurer.

Mr. Glaisher said :

" The first appearance of the balloon as a means of ascending into the upper regions of the atmosphere had been almost within the recollection of men now living, but, with the exception of some of the early experiments, it has scarcely occupied the attention of scientific men ; nor had the subject been properly recognised as a distinct branch of science.

" The main reason for this may have been that from the very commencement, balloons have been with but few exceptions employed merely for exhibition or for the purpose of public entertainment.

" A chief branch of inquiry by the Society would be that relating to mechanical invention, for facilitating aërial navigation and obtaining a change of locality at the will of the aëronaut, all contrivances for this purpose having hitherto failed.

" When we consider that the act of flying is not a vital condition, but purely a mechanical action, and that the animal creation furnishes us with models of every size and form, both simple and compound wings, from the minutest microscopic insect to the bird that soars for hours above the highest mountain range, it seems remarkable that no correct demonstration has ever been given of the combined principles upon which flight is performed, or of the absolute force required to maintain that flight, and it would be the office of the Society to bring forward any information or successful experiment illustrative of a theory."

And so the Society was established.

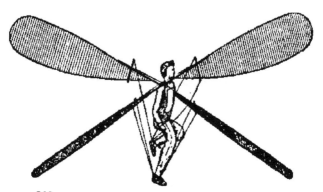

1866.—INVENTION OF BOURCART. FOUR BLADE-LIKE WINGS WORKED BY PEDALLING.

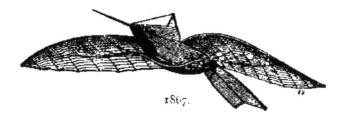

1867.

THE CHICKEN BONE.

1868.—M. Tissandier and a professional aëronaut made a voyage over the North Sea in a balloon called the " Neptune."

The machine made a splendid ascent, and was soon floating in mid-air buoyant as a feather, at a height of 4,000 feet, bound, as they fondly hoped, for the coast of England ; but they were carried out to sea, floating like gossamer into the very heart of cloudland.

Thus hovering miles above the world and its common-place cares, they enjoyed an interval of transcendent delight, rudely broken in upon by the professional aëronaut, a creature of appetite, who pulled the valve-rope unbidden, thus causing them to descend from their cloudy paradise into the grosser atmosphere that immediately surrounds the earth, where they at length bethought themselves of lunch. M. Tissandier finished his portion of a fowl by tossing a well-picked drumstick overboard. For this imprudence the professional was down upon him immediately. "Do you know," said he, "that to throw out ballast without orders is a very serious crime in a balloon?"

M. Tissandier was at first inclined to argue the point, but on looking at the sensitive barometer he was fain to admit that in consequence of the disappearance of the chicken-bone the balloon had made an upward bound of nearly thirty yards.

After a pleasant voyage, they sighted a lighthouse and descended near the spot where Rosier fell and was killed in 1785.

THE SIEGE OF PARIS.

A STRANGE VOYAGE IN A BALLOON.

On the 24th November 1870 M. Paul Rolier, the eminent engineer, with a rifleman ascended in a balloon in order to convey dispatches from

Paris to the exiled party at Tours. On taking his place at the railway station Du Nord in the balloon, M. Rolier conveyed 250 kilograms weight of despatches and six pigeons.

The Prussians having surveyed the sky as they were blockading the city, and had cannons cast for the purpose of directing them against balloons, it was necessary to set out at night in order to avoid the German projectiles.

Scarcely had the balloon started than it rose to a height of 2,000 metres, challenging the whole Prussian Camp, whose fires M. Rolier perceived like the phosphorescence of a considerable number of glow-worms. The wind blew very violently, the balloon was rushing at a fearful rate, and presently they heard the noise of waves breaking on the shore. All at once a thick fog enveloped them; suddenly the fog cleared and they found they were proceeding to sea. Eighteen vessels were in sight, and seeing a small French corvette with a tri-coloured flag, they let the balloon fall down into the sea and wait there till the French vessel should take them up; signals were made but unperceived, and the corvette disappears.

Then a German vessel fires a cannonade and misses the balloon; then a vertical current carries them off.

Again they descend and the waves cover the balloon with foam, the rope is moistened, which retards their course. M. Rolier casts into the sea a sack of papers and letters, again they rise and go eastward towards the open Polar Sea—

now they proceed towards firm land and graze the tops of trees.

By the aid of the rope hanging out M. Rolier at the risk of breaking his back, descended in safety with his companion.

The balloon immediately rises, almost suffocating them with the escaping gas; they were exhausted and were fainting.

They rise; where are they? In the snow; they see wolves; they walk five or six hours in the silence of the snowy solitude.

At length they take shelter in a hut, a little coal is still burning; it was a woodman's hut; and soon after two men arrived, but they could not understand one another, but one of them drew from his pocket a box of matches: M. Rolier took it, looked at it, and read "Christiana"; they were in Norway.

The woodmen conducted them to a small village where was a minister, a doctor, and an engineer of mines, named Neilson, who spoke French very well.

Jules Verne has never related a more extraordinary journey.

At Drammen they found their post bags, and the pigeons (still alive), a barometer, a sextant, and other objects.

M. Rolier had happily preserved two rolls of one thousand francs. He changed these for Norwegian money; and each one of these pieces of twenty francs was afterwards sold for nearly one hundred francs, and preserved there as memorial medals; they also made of the grappling

irons medals in commemoration of the event,
which were eagerly sought for in Norway.

At an entertainment in Christiana given by
a club of officers, M. Rolier collected 15,000
francs, which he preserved for the wounded
French soldiers, he exhibited the balloon and
added the product received to the above; so
that from this strange voyage the engineer Rolier
brought back from Norway to France more than
47,000 francs, which was deposited for the relief
of the wounded soldiers.

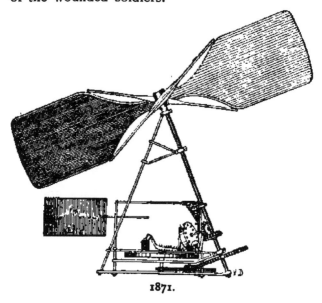

1871.

CHAPTER VIII.

1871 to 1885.

———

"A journey more pleasant, as safe and more soon,
We can take when equipp'd with a full-blown balloon."

BALLOON POST.

1871.—During the siege of Paris the beleaguered city contrived to communicate with the outside world by balloon, and the Government entered into a contract with M. Nadar for the despatch of a series of them under the direction of experienced aëronauts.

The first postal balloon, the " Neptune," was freighted with reports, letters and dispatches on its voyage on the morning of the 23rd September, from the Place de Saint-Pierre at Montmartre, and accomplished its mission successfully, as was ascertained by the return of the carrier pigeons it had taken out.

The Figaro of the time published the following :—

"What a strange fellow is Nadar !
Photographer and aëronaut !
He is as clever as Godard.
What a strange fellow is Nadar !

Although, between ourselves, as far
As arts concerned, he knoweth naught.
What a strange fellow is Nadar!
Philosopher and aëronaut.

At Ferrières, above the park, behold him
    darting through the sky
Soaring to heaven like a lark,
Whilst William whispers to Bismarck:
'Silence! see Nadar there on high
At Ferrières above the park!
Behold him darting through the sky!'"

AN UNINTENTIONAL JOURNEY.

March 17th, 1885.

In the *Times* of this date is the following:—
"Yesterday morning some labourers at work
in a field near Bromley were considerably
astonished to see a large balloon bounding
across some fields a short distance from them.
It was at length brought to a standstill by the
grappling-iron getting entangled in a tree. The
men proceeded to the assistance of the aëronauts,
who proved to be M. Ferdinand Dubois of the
Société aëronautique of Paris, and a Belgian
gentleman named Farenza.

"They had undergone a perilous balloon
adventure, having crossed the Channel much
against their will, and quite contrary to their
original intentions.

"The ascent was made on Saturday, a few
miles from Antwerp, the aëronauts intending if

possible, to descend somewhere near Brussels. All went well until M. Dubois throwing out ballast, they ascended higher. Coming into contact with a fresh current of air, they were carried in a contrary direction, and at nightfall were carried rapidly out to sea.

"The aëronauts, naturally much alarmed, endeavoured to attract the attention of some vessels they saw beneath them. Failing in this, M. Dubois deemed it prudent to throw out more ballast, so as to secure as high an ascent as was deemed advisable under the circumstances.

"All the provisions they were provided with were some sandwiches, biscuits, two flasks of brandy, and some water. These they utilised to the best advantage, and when morning dawned they found themselves far away out at sea.

"For the greater part of Sunday they were over the sea, but, as dusk set in, they were delighted to find themselves being carried toward land, and found themselves passing rapidly over a town, which they believed to be Folkestone, and they descended, as has been stated, near Bromley."

CHAPTER IX.

1885 to 1901.

———•——

1885–6. *From "La Nature."*

NAVIGABLE BALLOONS.

M. Renard, captain of the *Chalais Meudon* navigable balloon, presented to the French Academy of Science a report of the experiments made with that balloon in 1885.

On August 25th the balloon, which had been already filled for a certain time, having lost a considerable portion of its ascending force, M. Renard was under the necessity, on this occasion, of dispensing with the services of a third aëronaut, and mounted in the company solely of his brother, Captain Paul Renard. The wind blew from the east, and the speed, measured at a low height by means of small balloons, appeared to be no more than five metres a second, but were unable to gain the aërial current which prevailed at 250 metres above the valley of Chalais Deswug.

Nevertheless, to continue the experiment, and fearing to be carried away above the woods of the Chaville quarter, M. Renard turned the head of the balloon a little to the right, and soon, under the combined action of the wind and its own speed, it took a southern direction, and the backward movement continuing, alighted after a voyage of 50 minutes close by the farm of Villa Coublay, whither he had directed it.

The second definite experiment did not come off till September 22nd, when the wind was blowing from N.N.E.—that is, from Paris—and its velocity in the lower strata varied from 3 to 3·50 metres per second.

This time the aëronauts had their full complement of three: Captain Renard at the helm and the motory machine, Captain Paul Renard taking measurements and various observations, and, in addition, M. Dulé-Poiterin. They started at 4.25 p.m. in a moist and foggy atmosphere. The spiral was set in motion and the head directed towards Paris.

Though at first inclined to yaw, the course of the balloon soon righted itself, and crossing the railway line above the station at 4.55, the balloon reached the Seine towards the western extremity of the island of Billancourt at 5 o'clock.

The turning of the balloon was easily effected, and, aided this time by the aërial current, it approached its point of departure with surprising rapidity, and in ten minutes the little skiff touched the sward, whence it had ascended.''

Out of seven voyages, from August 9th, 1884, to September 23rd, 1885, the aërostat has in five returned to its point of departure.

In 1899–1900 M. Deutsch offered a prize of
£4,000 to the first balloonist who should start
from some fixed point in Paris, make the circuit
of the Eiffel Tower, and return to the place
whence he came without touching the earth, in
the space of thirty minutes.

M. de Santos Dumont won this prize with
his balloon, which was cigar-shaped and measured
120 feet from tip to tail, and held a motor of
sixteen horse power which drove a propeller, and
was guided by a rudder of silk stretched over
a bamboo frame.

With this machine he travelled against the
wind a distance of two miles, made the circuit of
the tower, and was wafted back to the starting
place.

November 17th, 1901.

AN EXCITING ASCENT.

Mr. Spencer, Mr. Bacon and his daughter
ascended from Newbury Gas Works in a 56,000
cubic feet balloon about four in the morning.   In
a few minutes they entered into a tract of warm
steaming cloud where all was darkness, save
for the feeble glimmer of a Davy lamp, and
there they hung inert.

The cause of their lack of buoyancy was that
the balloon had been filled many hours before.
Exposure had chilled it, and its vast surface
was heavily weighted with dew ; and it was only
by a large expenditure of ballast that they were

able to rise through the cloud into the clear air above. More than once the cold sent them down again, and yet more sand was discharged.

At last they were able to maintain themselves at an altitude of 40,000 feet. It was bitterly cold, and the balloon mysteriously ascended, and then about six o'clock they heard, from beneath the cheerless 1,500 feet of cloud, the prolonged crowing of cocks.

At this point they became aware of an alarming change in their circumstances. The balloon began to ascend, and it continued ascending into a heaven growing every minute warmer. They had parted with a weight of ballast which had kept them at a lower level ; the dead weight of moisture on the surface of the silk was evaporating, and the rays of the sun were warming the gas, and thus accelerating their upward flight.

Why not open the valve and let out some of the gas? As fortune would have it, to prevent leakage during the long hours of waiting, the balloon had been equipped with a "solid valve." This valve could only be used once, and at a height of 5,000 feet it would have been sheer madness to tear a gaping chasm two or three feet across in the top of the balloon ; there was no alternative but to wait and hope for the best. The balloon was rising to an altitude of 10,000 feet in the burning sky and drifting westward over an unseen country towards the sea.

In the earlier part of the morning our balloon had been apparently haunted by a ghostly visitor. We were then some 100 feet above the cloud-floor, which, however, seemed almost at our feet, with a surface which looked as compact and hard as an ice-field, and we heard repeatedly the sound as of someone walking softly outside. The explanation of these mysterious sounds was presently traced to the moist shrunken netting giving out under the hot sun and yielding now and again with gentle release to the expanding gas.

Just as their straits seemed desperate, a cold, strong wind from the west struck them, diverted their course, and caused the balloon to descend in spite of the mid-day heat of the sun. When at last they came crashing to earth in a fierce gust which hurled them through barbed wire fencing, hedge, and oak tree, they had had a run of ten hours, the incidents of which, one fancies, will furnish them—and especially the lady, who broke her arm in the "landing"—with sufficient material for nightmares to last them a lifetime.—"*By Land and Sky*," by John M. Bacon, M.A., F.R.A.

TWO CURIOUS ITEMS.

(I) THE SACRED GLORY.

In *Travels in the Air*, by James Glaisher, F.R.S., we read :—

"Our height was never over 2,300 feet.

"About 6.45 the shadow of the balloon became white, as I had previously noticed in our morning ascent over the Loire.

"On examining attentively this phenomenon, I found that it is really due to the reflection of the solar rays from the dew-drops on the grass of the fields, or the leaves of the trees, and that this occurs either in the morning or in the evening.

"When the motion of the balloon carried its shadow over the Seine the latter became quite invisible. On the wood of St. Germain it appeared as an immense white aureola, the centre of which was occupied by a dark circle.

"I have received several curious letters with respect to this shadow, one of which, from a medical gentleman, and another from a gardener, attribute the phenomenon rightly to the dampness of the soil. The latter stated that if I had ever walked early in the morning over the ground covered with dew, I could scarcely have failed to observe the shadow of my head surrounded by a *sacred glory.*"

(2) "YOUR PASSPORTS, GENTLEMEN!"

"After having crossed the Seine, we glided along at a very slight distance above the ground, and were suddenly saluted from below with the words: "*Your passports, Gentlemen!*" from two gendarmes galloping along the road to St. Germain. As there was a very good reason for our not throwing down our passports to them, Goddard begged them to step up and verify them, emptying out a bag of ballast as he did so.

"The two police agents doubtless thought upon the modification that would have to be introduced into the mounted police force as aërial navigation comes more into vogue."

BOILING BALLOONS.

In an article, "A Trip Heavenward: Ballooning as a Sport," in *Blackwood's Magazine*, Vol. 158, Major B. Baden - Powell tells the following story :—

"Some years ago I bought a balloon from a professional aëronaut. Though not a new one, it was sound enough for my purpose. I used it for several ascents, knocked it about a good deal, had it patched and altered, and finally stowed it away for some months in a cellar.

"After that I knew it was probably not trustworthy, and therefore determined to get rid of it. I asked the maker to buy it back as so much old material. Of course, he was only too ready to agree that it was utterly worthless as anything else, and so I parted with it for a few pounds.

"Some time afterwards I happened to hear that this very balloon had made another ascent, and I therefore ventured to remind the aëronaut of the exact nature of our transaction, recalling the fact that the price he paid me was not the market value of a serviceable balloon, and I presumed to advise him against the risk of trusting his life and limbs to such utterly worthless old material as he had described it.

"Then, it seems, he took the matter to heart, and, like me, thought it best to be rid of the thing, so he sold it to another professional named Dale. This man emulating the magician in Aladdin, had a great invention for converting old balloons to new ones.

"He took the old 'Eclipse' and put it in the pot, and boiled it down with soda and other chemicals, till all the varnish had disappeared, and left a mass of snow white cambric as clean as it was on the day it was born.

"He varnished the stuff afresh, and then turned out a splendid balloon, quite unrecognisable from the good old 'Eclipse' which had its name in 'life-sized' letters painted on it.

"Some years after a young Naval Reserve Officer in India became possessed of some idea with regard to balloons and parachutes for military purposes, and, with the idea of putting his theories to a test, sent home for a balloon. Dale had the very article for him, and shipped it off at once.

"Poor Mansfield made his first ascent at Bombay; but ere he had attained an elevation of 200 or 300 feet, the balloon burst asunder and fell to the ground, the unfortunate aëronaut being fatally injured.

"Meanwhile poor Dale doubtless thought he had found the elixir of life for balloons, and prepared a second old balloon in the same way, and, what proves that he did not realise the danger or intentionally commit so awful a blunder, made an ascent himself in it, accompanied by his son and others.

"This balloon acted in just the same way as the first, bursting ere it was clear of the Crystal Palace grounds and dashing to earth its human freight, Dale and one of his companions being killed, the other dreadfully injured."

CHAPTER X.

AËRONAUTICS IN THE TWENTIETH CENTURY.

By Major B. BADEN-POWELL.

During the few years of the twentieth century which have elapsed, more progress has been made in all branches of aëronautics than in the previous 100 years. As regards the simple balloon, though no actual improvement in the form of apparatus can be chronicled, almost all previous records have been beaten. The greatest height well recorded has been attained, the furthest distance travelled, the longest duration in mid-air, while the number of ascents made has been so great, both in England and abroad, as to constitute a big record per year.

Propelled balloons have also been introduced in such numbers and with such success as to beat all that has been done before, and they may now be said to be practical machines of war. Even man-lifting kites have been improved and record heights attained. Small kites for meteorological investigation have been sent up to heights unheard of a few years back, and, indeed, the same may be said of sounding balloons. But it is with the flying machine proper that still more important strides have been made. Not only have men for the first time been actually lifted off the ground by mechanical means, but quite long flights have, on occasion, been made. All this has been accomplished within seven years.

Unfortunately during this short period have to be chronicled the deaths of several of the persons most eminent in the history of aëronautics. Both Coxwell and Glaisher, who made such names for themselves on account of their great ascents for scientific purposes, died within a few years of one another. Professor S. P. Langley, who made such valuable experiments in aëro-dynamics, and who first succeeded in making a large model with engines to go for many hundred yards through the air; the Rev. J. M. Bacon, who probably made more ascents for scientific purpose than any other man. Colonel Renard, the inventor of the first airship to succeed in making prearranged return journeys, died recently in Paris. The names of Thomas Moy and F. J. Stringfellow, too, will ever be remembered in connection with the first attempts at making model aëroplanes and light engines.

It will perhaps be clearest to group the various incidents under their respective headings, rather than to relate what has been done in chrono-logical order.

F

FREE BALLOONS.

In the year 1900 was held the International
Exhibition in Paris, and in connection with this
an Aëronautic Congress was assembled at Vincennes. This included an exhibition of apparatus
of all kinds connected with the subject, but was
of especial interest owing to the prizes offered,
for the first time on so big a scale, for balloon
journeys. One hundred and fifty-eight ascents
were made, and that without any serious mishap.

On 16th September no less than 26 balloons
were inflated at once. The prize for altitude
was won by MM. Balsan and Louis Godard, who,
on 23rd September, attained a height of 8,558
metres.

On 9th October the world's record for distance
(also the prize for long duration) was attained by
Count Henry de la Vaulx and Count de Castillon.
The "Centaure" ascended at 5.20 p.m., and after
remaining up for two nights, and ascending to
a height of 5,200 metres, finally came down close
to the town of Korostichef, in the province of
Kiev, Russia. The voyage lasted 35 hours
45 minutes, and from point to point totalled a
distance of 1,922 kilometres (close upon 1,200
miles).

M. Jacques Faure on September 1st left
London in his balloon and returned to France by
night.

In August 1900 Captain Spelterini made a
novel kind of ascent, taking his balloon, together
with a fill of hydrogen compressed in steel tubes,

to the top of the Righi, whence he ascended with two companions, passed over the Alps, and descended on a mountain in the Canton of Glaris.

A fatal balloon accident occurred in Italy in July. Captain Venni, accompanied by Count di Montecupo and Signor Pellizoni, ascended from Naples and were at once carried out to sea, but hoped to land on the island of Capri. This, however, they missed, and descended in the sea, hoping to be picked up. After five hours of clinging to the ropes, Venni and Pellizoni let go and were drowned; the other passenger hung on for another six hours, when he was rescued by a fishing boat.

On July 31st, 1901, Dr. Berson and Dr. Süring ascended from Berlin and reached a height of 34,400 feet. Although both aëronauts lost consciousness at 33,600 feet, yet the self-registering instruments record the former height. It will be remembered that Messrs. Glaisher and Coxwell ascended on 5th September, 1862, to a height which they had reason to estimate at 37,000 feet. This was by far the greatest height ever attained by man, although there was no absolute certainty about the exact altitude.

On April 5th, 1906, the record journey for duration was commenced. Dr. Kurt Wegener, of the German Aëronautical Observatory at Lindenberg, and his brother Dr. Alfred Wegener, made an ascent at 9 a.m. on that day from near Berlin, in a balloon of 42,000 cubic feet. Passing over Wismar, they crossed a part of the Baltic

F 2

and travelled right up Denmark, and just when it seemed probable that they would cross over to Sweden the wind changed and drove the balloon directly south, and, passing over Kiel at six o'clock on the second evening, the journey was continued down through Germany to Aschaffenburg, where a descent was made at 1.30 p.m. on April 7th, the journey thus having lasted fifty-two and a half hours.

An attempt to cross the Mediterranean was made by Count de la Vaulx in 1901. The balloon was provided with a large trail rope fitted with "deviators." Starting from Toulon on October 12th, the balloon was carried in a southerly direction, but the wind soon changed, and eventually, after having been forty-one hours in the air, descended to the cruiser which had been told off to accompany it.

Balloon racing, although not unknown in former years, has recently become very popular.

Besides the many events in Paris, in England the Aëro Club has organised several contests. In Germany some big events have taken place. In Italy there were several such contests in connection with the Milan Exhibition in 1906. But the greatest of all, from a popular point of view, was the race for the Gordon-Bennett Cup, starting from Paris on September 30th, 1906. Sixteen large balloons took part in this race, Great Britain, France, Germany, and Spain having three each, while America had two, Italy and Belgium one each. Seven of the balloons crossed over the Channel to England, the first place being awarded to Lieut. Lahm, representing the United States, who descended near Whitby in Yorkshire.

In November 1906, Mr. Leslie Bucknall and Mr. P. Spencer journeyed in a balloon from London to Nevy in Switzerland, a distance over 400 miles.

On April 11th, 1907, Dr. Wegener and Mr. Kock, starting from Bitterfeld, near Berlin, passed across Holland and over the North Sea and landed at Enderby, Leicestershire, a distance of 812 miles, in nineteen hours.

SOUNDING BALLOONS.

Great progress has been made during the last few years in the matter of sounding the upper strata of the air with small balloons carrying self-recording instruments. Already in 1895 such a balloon had been sent up to a height of over $13\frac{1}{2}$ miles. But not much more had been accomplished until the founding of the International Commission for Scientific Aëronautics in 1900. This institution was organised to carry out ascents with balloons, kites, or other means of observing the condition of the upper air on pre-arranged days, so that the results could be compared. Representatives from all countries were appointed, and certain dates fixed, usually once a month, on which the observations were to be made. A large number of ascents by sounding balloons have been carried out, usually in about 17 or 18 different places in

Europe and America at the same time.   Heights
up to 20,000 metres have often been attained, but
the record was obtained in the ascent of a rubber
balloon from Strassburg on 3rd of August, 1905,
which reached 25,800 metres, or over 16 miles.
It seems remarkable that we are now able to
accomplish this, probing our atmosphere to such
an enormous height, far above where, owing to
the extreme rarefaction of the air, human life
would be possible.

KITES.

But the subject of sounding balloons cannot
be separated from that of sounding kites, for the
latter are now coming to be considered as an
important adjunct to meteorological observation.
Though kites have in years gone by often been
used in this connection, notably by Benjamin
Franklin, to test atmospheric electricity, yet it
is only of late that great heights and regular
observations have been carried out.   By letting
up a kite on a thin steel wire, and, when it can
lift no more line, attaching a second kite to the
wire, a very great altitude can be attained.   At
the Blue Hill Observatory in America, Mr.
L. Rotch first introduced kites as a regular part
of the equipment, and to-day there are many
observatories so furnished.

In 1904, M. Teisserenc de Bort, who has for
many years devoted himself to kite-flying for
meteorological study, organised a cruise on the
Baltic in a Danish gunboat.   Almost daily flights

were made to very considerable heights, culminating on April 25th with a record altitude of 19,360 feet. This required a wire of 38,000 feet in length.

One of the highest kite flights was carried out by Dr. Assman at the Lindenberg Observatory in 1906, when, with a series of six kites having a total area of 323 square feet, a height of four miles was reached. At the highest point, where the temperature was 13° F., the wind velocity was 56 miles an hour, though only 18 miles per hour near the ground. Other high flights were made in January 1906, when 18,000 feet was attained at Hamburg and 15,000 at Kiel.

On June 25th, 1903, an International Kite Competition was inaugurated by the Aëronautical Society of Great Britain and took place near Findon, Sussex. Eight large kites were entered, but unfortunately the wind was not favourable to any great heights being obtained, the prize being offered for the highest flight over 3,000 feet. Mr. C. Brogden's six-winged kite took the first place, though only attaining a height of 1,800 feet.

At the St. Louis Exhibition a year later another such contest took place, a box kite of ordinary pattern taking the prize for stability, but the high flights were here also failures on account of lack of wind.

Kites have also come into vogue for other purposes than meteorological observation. Man-lifting war-kites have now become a regular item in the material of His Majesty's forces. At.

Aldershot much practice has been carried out since the first trials in 1896.   In 1905 a man was lifted to a height of 2,600 feet, and remained there for an hour, but since then several ascents have been made up to 3,000 feet.   At this height a man is almost invisible, and is practically beyond the range of rifle shots, so that this is an ideal position for an observer to watch the movements of an enemy.

AIRSHIPS, OR PROPELLED BALLOONS.

Balloons propelled by screws driven by engines, have made great strides during the period under review.   It is often forgotten that a spindle-shaped balloon carrying a steam engine to work a screw-propeller, made an ascent as long ago as 1852, and that the first such machine (" La France ") to perform journeys to and fro, returning to its starting point, was made 25 years ago.

It was in the spring of 1900 that M. Deutsch de la Meurthe offered his prize of 100,000 francs (£4,000) for the first dirigible airship that should go from the Aëro Club grounds at St. Cloud around the Eiffel Tower and back within half an hour, a total distance of close on seven miles. This distance, it may be noted, was not so great as that which had been twice successfully covered by the airship " La France " in 1885.

M. Santos Dumont had already built three airships before the close of the nineteenth century,

but all had been entire failures. In August 1900,
his "No. 4" was complete. This was 129 feet
long by 17 feet at its greatest diameter, and
contained nearly 15,000 cubic feet. It was fitted
with a 7 h.p. engine and a screw propeller of
13 feet across, which was placed in front of the
car, or rather pole, on which the aëronaut's seat
was fixed. This machine also did not quite fulfil
all that was expected of it, so that next year it
was enlarged, a girder "keel" suspended to it
by steel wires (hemp rope having been previously
used) and a 12 h.p. 4-cylinder motor applied.
. The propeller was placed in the rear. With this
vessel a satisfactory round trip was made, and
later on in an attempt to accomplish the course
of the Deutsch prize the balloon collapsed after
rounding the Eiffel Tower and fell on the roof
of an hotel.

"No. 6" soon followed. It was 110 feet long
and 20 feet in diameter, and contained 23,200
cubic feet. The engine was of 12 horse-power
water cooled. With this machine, on the 19th
October 1901, Santos Dumont won the prize,
oddly enough covering the distance within a
minute or two of the requisite time. After this
M. Santos Dumont built several more airships
with which he had more or less success, but
none accomplished a greater speed or distance.

Thus, although Santos Dumont had accomplished nothing very extraordinary, yet he
rendered a great service in bringing the subject
to the fore, and creating so much public interest
in aërial navigation.

During the next few years there was a perfect craze for airships, and a number were constructed. In Paris, a huge twin vessel was built for M. Rose, but it was never even able to ascend. Señor Severo, of Brazil, then had a great machine built, called the "Pax," which made an ascent in Paris, but soon after the start it somehow caught fire and the whole crashed to earth, the two occupants being instantly killed. This was in May 1902, and the year following another unfortunate disaster occurred. Baron de Bradsky had an airship made which had a long girder car suspended by wires from the balloon. These wires do not seem to have been made strong enough, and in the first ascent on October 13th, after the balloon, which had been unable to stem the wind, had drifted across Paris, the wires all broke and the car fell to the ground separated from the balloon. The inventor and his mechanic, Morin, were dashed to pieces.

Other airships have since been made in Paris, notably a huge vessel for M. Deutsch, which so far has not had much success. More recently one has been made for Count de la Vaulx, which is a comparatively small machine of some 26,000 cubic feet, and is provided with a 16 h.p. Ader engine. Quite a number of ascents have now been made with this machine, all with success, a speed of about 13 to 14 miles an hour being attained.

Of all the airships hitherto built perhaps the most successful are those known by the name of

"Lebaudy." M. Henri Julliot, an engineer in the sugar factory of the MM. Lebaudy in Paris, designed these, of which the first was built in 1903, and made its first ascent in May of that year. This machine made some sixty successful journeys, always (with one exception) returning to its shed at Moisson. Finally this vessel went to Paris and thence to Chalais-Meudon, where it was to be tested by the military authorities, but, unfortunately, it was wrecked on its descent, being dashed among the trees. A second balloon was made to practically the same pattern, which proved also quite a success, but this also was wrecked when on the ground undergoing military trials. However, so satisfactory were these machines considered that the French Government decided to get one made by the same constructors and the "Patrie" is the result. This balloon is of 108,000 cubic feet capacity, has engines of 75 horse-power, and has proved capable of going 28 miles an hour, carrying a crew of five men.

Count Ferdinand von Zeppelin, a lieutenant general in the Würtemburg army, had for many years made a study of aëronautics, and conceived the notion that a very long cylindrical balloon could be propelled rapidly through the air.

It has often been suggested that the best means of obtaining rapidity, so essential a feature in any balloon that has to be propelled through the air, is to keep it distended by a rigid framework. Mansfield, in his delightful book on "Aërial Navigation," discussed this matter very

thoroughly. Once one adopts a girder-like framework, there is no difficulty in having a balloon
of very great length. If resistance is to be
measured by the area of the midship section,
then the lengthening of the vessel will not
materially increase the resistance. On the other
hand, a framework such as this is bound to be
comparatively heavy. But here another principle
comes in, which is that the larger the volume
of a balloon the greater the lift, while the surface
area and the head resistance do not increase in
so great a proportion. Hence a very large
balloon is more suitable for propulsion than a
small one, a framework would be impossible
without very great size, but the larger the vessel
the greater is the available lift. Zeppelin, therefore, decided on making a huge vessel. This
was built in a large shed supported by pontoons
floating on the lake of Constance. A vast framework of aluminium girders trussed with wires
and netting was made in the form of a cylinder
with conical ends. It was 420 feet long and
37 feet in diameter. The framework was divided
into a number of compartments, into each of
which a separate balloon was placed, and a
smooth outer skin enclosed the whole. The
propelling apparatus consisted of four screws,
two being placed on each side of the vessel,
near the bow and near the stern, and on a level
with the axis of the balloon. Two sets of
engines were employed, each originally being
of 15 horse-power. The first ascent took place
early in 1900.

This was hardly a success, and an accident which happened shortly after caused a strain on the whole framework. The immense expense of these experiments exhausted the available funds for some time, but more money was raised from various sources, and at last, in 1906, another trial was made, which proved most successful. Various improvements had in the meantime been made, especially as regards the engines, the original ones having been replaced by two 85 horse-power Daimler motors. On October 9th the ascent was made and the airship cruised around Lake Constance, manœuvring for two hours. The average speed during this time was twelve metres per second (over 29 miles an hour).

In May 1906 an airship invented by Major Parseval was tried near Berlin. It was said to be provided with a 90 horse-power Mercédès motor, and that it made several short return journeys with complete success.

In England, not much has been done with navigable balloons, probably our windy climate not being favourable to success. However, the late Mr. Stanley Spencer made one with which he made some sensation by crossing over London, though only with the wind. Later he made one or two short return journeys. Mr. Beedle made another balloon which did not, however, succeed, and finally Dr. Barton built an enormous machine of 200,000 cubic feet capacity. This was fitted with two 50 horse-power engines driving four propellers. But its first voyage was its last, for it was quite unable to stem the light wind blowing, and the elaborate structure of bamboo which formed the car was smashed to pieces on descending.

In America several dirigible balloons were constructed to take part in the competitions at the St. Louis Exhibition, where prizes to the amount of £40,000 were offered. The Benbow airship was peculiar in being propelled by feathering paddle-wheels. But its speed proved to be insignificant. The François - Lambert machine did no better, but Mr. Baldwin's "Arrow," though not attaining any great speed, and usually being carried away by the wind, did, on several occasions when very calm, return to its starting point.

A large airship, 164 feet long, and containing 224,000 cubic feet of hydrogen, was constructed in Paris for Mr. Wellman to make an attempt to reach the North Pole. The balloon, with its two motors, one of 55 horse-power and the other of 25 horse-power, was taken to Spitzbergen, and underwent trials during the summer of 1906. But these not proving satisfactory, the apparatus was bought back to France, with the idea of making improvements and having another try next summer.

Early in 1907 Count de la Vaulx had a small dirigible airship constructed of some 26,000 cubic feet. It was provided with an engine of 16 horse-power working a single screw. This made a number of successful journeys, attaining a speed of nearly 22 miles an hour.

FLYING MACHINES.

The first event of importance in this line during the twentieth century was the trial of a machine designed by Herr Kress in Austria. The apparatus consisted of a long boat or punt above which were arranged three sets of wings. Vertical and horizontal rudders were placed behind, and the whole was propelled by two "resilient sail" screws near the middle, actuated by a 35 horse-power Mercédès motor.

In October 1901 the crucial test was made on a reservoir at Tullenbacher, near Vienna. The machine was driven along at a considerable speed, skimming along the surface of the water. Then it suddenly rose into the air, but immediately toppled over, and fell into the water, and sank to the bottom.

Some fifteen or twenty years ago the late Prof. S. P. Langley, of the Smithsonian Institution, Washington, commenced a series of elaborate scientific experiments to determine the resistance of the air to various surfaces driven through it, and the reactions resulting. After the results of them had been published, the United States Government decided on expending a certain sum to see whether a machine of practical military utility could be constructed, and £10,000 was voted to Prof. Langley to carry out the trials. He thereupon built a man-carrying machine, almost exactly on the model of his smaller "Aërodrome" which had flown so successfully. To get a really suitable and reliable motor proved to be the most difficult task, and nearly three years elapsed before such was finally obtained. By the end of 1901, however, a 5-cylinder motor of 52 horse-power, weighing with cooling water and all accessories, under 5 lbs. per horse-power, was in running order. The machine complete, with aëronaut and engines, weighed 830 lbs., and had a sustaining surface of 1,040 square feet. Many more delays, however, occurred, and it was two years before everything was in order. In the meantime a quarter-sized model had been constructed. This weighed 58 pounds, had a sustaining surface of 66 square feet, and was provided with a motor of $2\frac{1}{2}$ to 3 horse-power. The model worked perfectly, flying directly ahead on an even keel.

Finally, on 7th October 1903 a trial was made. Most unfortunately, after all this lengthy preparation, a portion of the apparatus caught in the launching-ways, and caused the machine to dive forward and fall into the water. This necessitated many repairs, and a second trial could not be attempted till two months later. Again the machine was upset on launching, and winter prevented further trials at the time. But the funds had become exhausted, and the authorities were loth to devote more to such experiments.

By far the most important steps made up to now in connection with aërial navigation by gasless machines have been the trials of apparatus by two brothers, Orville and Wilbur Wright, of

Dayton, U.S.A. They had for some years been practising in a gliding machine, after the manner of Lilienthal and Pilcher. Their apparatus was a modification of that designed by Chanute, and consisted in two superposed planes, capable of being twisted or warped, so as to balance and steer the machine without shifting the weight of the man, as had hitherto been the practice in such apparatus. Another novelty was that the operator was lying prone, instead of being upright, which had great advantages in that less resistance was offered to the air, and injury from shock on landing less liable to occur. By such means large surfaces could be controlled and managed, and their 1902 machine was of 300 square feet surface. After much practice and experience had been gained in the management of these machines in mid-air, the experimenters decided on applying a motor and propeller to the apparatus, and on December 17th, 1903, a successful effort was made to rise from the ground against the wind. Other flights followed, the longest lasting 59 seconds, and covering a distance of nearly 300 yards. Further trials were made during the next year, and many improvements effected. In September 1905 the machine rose, and was steered round and round a field, the distance travelled being calculated at 11 miles; and on the 5th October a flight was made lasting over 30 minutes, and covering over 24 miles of the course.

As the inventors have kept all details secret, it is not possible at present to give any further

G

description of the apparatus actually employed, though many unauthentic (and possibly incorrect) facts have been published, to the effect that the general arrangement is the same apparatus as used by these experimenters in gliding, but with a motor and screw propeller attached.

Herr Ellehammer in Denmark has for some years been testing a machine with two wing-like superposed planes, which is said to have made successful flights of 50 metres.

During 1906 and 1907 quite a number of more or less successful trials of aëroplane machines were made in the neighbourhood of Paris.

Early in 1906 M. Santos Dumont, having failed to get satisfactory results from an apparatus he had made on the *helicoptere* system, decided to try one on the aëroplane principle. This consisted of two superposed planes forming a slight dihedral angle, and divided into compartments by vertical divisions. It was 12 metres across, and the total lifting surface was 80 square metres. Towards the front, at the end of a square tube covered with canvas, was arranged a box-like rudder. The whole machine, with aëronaut, weighed about 550 lbs., and it was driven by a 50 horse-power engine rotating an aluminium screw propeller, 6 feet diameter in rear of the machine. On September 13th the machine was tested, and after a preliminary run it was successful in rising off the ground and accomplishing a short flight of a dozen yards. Then, however, it fell heavily and was damaged. By October 23rd it was repaired and again tried.

After a run along the ground of some 80 yards at a speed of 30 miles an hour, it rose steadily to a height of about six feet, and continued travelling in the air for 60 metres. On November 12th another trial was made, when the machine rose to a height of about 16 feet, and travelled a distance of 220 metres without touching the ground, the flight lasting $21\frac{1}{5}$ seconds, but then the apparatus again fell heavily to the ground and was damaged. In the following year M. Dumont built another machine of a different pattern. In this the planes were made of thin panels of mahogany instead of canvas stretched on a bamboo framing. The rudder arrangement was now placed in rear of the machine, and the propeller in front. A number of trials were made, but all so far have ended in failure to rise off the ground. A trial of his last year's machine also proved disastrous, resulting in a smash up.

M. Bleriot, who first tried a machine somewhat like a large box kite mounted on catamarans, to float on the water on Lake Engheim, finally developed an apparatus looking rather like a large bird with narrow outstretched wings. It was of 13 square metres surface, and weighed $5\frac{1}{2}$ cwt. With this, in April 1907, he succeeded in rising off the ground and travelling for about 10 yards.

The apparatus of M. Vuia consisted of a pair of large outstretched wings with an area of 20 square metres, below which was a framework to carry the man and motor and propeller, the whole being mounted on four bicycle wheels. The motor, actuated by liquid carbonic acid, was capable of developing 25 horse-power. After several unsuccessful attempts, the machine was again tried at Bagatelle in December 1906, and after getting up a speed of about 40 kilos. an hour, it just lifted clear of the ground, and accomplished a free flight of about 6 metres. In the March following it again made some successful jumps or flights up to 50 feet, rising 4 feet from the ground.

M. Delagrange's aëroplane machine has a lifting surface of 50 square metres, the main plane being 32½ by 6½ feet, rear planes 16½ by 6½ feet, and is propelled by a 50 horse-power Antoinette engine. It has a fore rudder somewhat similar to that used in Santos Dumont's first machine. After several unsuccessful attempts and mishaps, this machine rose from the ground in March 1907 and flew for about 50 yards.

Two other machines of a somewhat similar kind have recently been constructed in Paris. One, designed by Captain Ferber for the Antoinette Company, is not unlike that of M. Bleriot, having a pair of single surface wings of 25 square metres. It is to be driven by a 100 horse-power motor, actuating a propeller in rear.

The Kapferer aëroplane has two wide superposed main planes 11 metres in span with smaller ones both in front and behind, the lifting surface totalling 25 square metres. It is driven by a 25 horse-power Buchet engine.

In connection with flying machines must be recounted a sad fatality which occurred in America in July 1905. Prof. J. J. Montgomery, of Santa Clare College, California, had designed a large gliding machine, which was, on several occasions carried up to a height of several thousand feet by a balloon, when the operator, J. M. Maloney, detached it and descended, making various evolutions, and guiding the machine hither and thither as it slowly glided to earth. On the last occasion, however, something went wrong with the apparatus when at a great height; the aëronaut lost control, and was dashed to the earth and killed.

Such, then, are the main points of the history of aëronautics of recent years. As the progress has been so remarkable each year, it looks as though we should have still further leaps of progress in the near future, and it becomes indeed difficult to foresee the vast changes that may occur within the next few years affecting war, commerce, and our everyday life.

AËRIAL POEMS

AND QUOTATIONS.

1784.

Wretched man ! what food
Will be conveyed up thither to sustain
Himself and his rash 'army, where thin air
Above the clouds will rise, his entrails gross,
And famish him of breath, if not of bread ?

Milton's " Paradise Lost."

A song in the praise of balloons I will give,
    Since nothing but them I find will go down ;
Lunardi, who first took an aërial voyage,
    My theme he shall be, for he gained much
        renown.

CHORUS—

    Lunardi's balloon rose in the amain,
    Turn'd about, took a route, safely came down
        again ;
    The sky being clear, a fine sight we did obtain,
    Success to Lunardi and his first balloon.

How few the worldly evils now I dread,
No more confined this narrow earth to tread!
Should fire or water spread destruction drear,
Or earthquake shake this sublunary sphere,
In air-balloons to distant realms I fly,
And leave the creeping world to sink and die.

The Air Balloon, 1784.

A puffing at the air-balloon
You now behold it filled quite soon.
The people stare to see it fly—
'Zooks!' 'tis got surprising high!
The man in the moon, but not asleep
Old Catafogo takes a peep;
The clowns are startled at the sight—
'Tis burst! and now it comes down quite.

1785.

Upon my life
I'll take my wife
A ride in a balloon;
And pray, Sir, why?
Eh! Madam, aye!
A fine thing a balloon.

ROSIÈRE'S EPITAPH.

" Sacred to thee, Rosière, this stone,
    Who first th' advent'rous art essay'd
To rule at will the swift balloon.
    Amid'st the ambient air displayed;
When from Death's store a cruel dart was sent
To make the Aerostant thy Monument."

[Free translation.]

FANCY IN NUBIBUS:

Or, THE POET IN THE CLOUDS.

O ! it is pleasant, with a heart at ease,
Just after sunset, or by moonlight skies,
To make the shifting clouds be what you please,
Or let the easily persuaded eyes
Own each quaint likeness issuing from the mould
Of a friend's fancy; or, with head bent low
And cheek aslant, see rivers flow of gold
'Twixt crimson banks: and then, a traveller, go
From mount to mount through Cloudland, gor-
    geous land !
Or, list'ning to the tide with closèd sight,
Be that blind bard, whom the Chian strand
By those deep sounds possessed with inward
    light,
Beheld the Iliad and the Odyssey
    Rise to the swelling of the voiceful sea.

Coleridge.

1802 to 1811.

·THE PARACHUTE.

Charles Dibdin wrote the following lines on parachutes and balloons :—

Draw near, I pray, nor what I sing
 To aught amiss impute;
'Tis on a most ingenious thing,
 Yclept, a parachute.
Kindly brought over late from France,
 In fashion whilst we sprawl,
To teach us, in life's giddy dance,
 To guard against a fall.

From France we all our fashions brought,
 And 'tis but fair to note
That those who have the poison taught,
 Should teach the antidote;
And lest in fancy's air balloon,
 We for assistance call,
'Twas kind from th' influence of the moon,
 To guard us 'gainst a fall.

Like them our fashions then correct,
 As far as follies reach;
But let them nothing else expect
 To Englishmen to teach;
Nor fondly think they can dispute
 With us fair freedom's ball:
Our Union's the true parachute
 To guard against a fall,

How to be cautious in this sort,
 We need not to be bid;
And yet we kindly thank them for 't,
 As much as if we did:
For disaffection long time, now   ·
 Thank Heav'n ! has ceas'd to bawl—
Experience well has taught us how
 To guard against a fall.

Then fellow-subjects, neighbours, friends,
 United be and true;
So shall you ne'er the private ends
 Of the world's empire rue;
So, by no foreign arts ensnar'd,
 Your freedom to enthral,
Shall good old England be prepar'd
 To guard against a fall.

1811.

THE BALLOON.

The *airy ship* at anchor rides:
Proudly she heaves her painted sides
 Impatient of delay.
And now her silken form expands,
She springs aloft, she bursts her bands,
 She floats upon her way.

How swift! for now I see her sail
High mounted on the viewless gale,
 And speeding up the sky:
And now a speck in ether tost,
A moment seen, a moment lost,
 She cheats my dazzled eye.

Bright wonder ! thee no flapping wing,
No labouring oar, no bounding spring,
    Urged on thy fleet career:
By native buoyancy impelled,
Thy easy flight was smoothly held
    Along the silent sphere.

No curling mist at closing light,
No meteor on the breast of night,
    No cloud at breezy dawn.
No leaf adown the summer tide
More effortless is seen to glide,
    Or shadow o'er the lawn.

Yet thee, e'en thee, the destined hour
Shall summon from thy airy tower
    Rapid in prone descent:
Methinks I see thee earthward borne
With flaccid sides that droop forlorn,
    The breath ethereal spent.

Thus daring Fancy's pen sublime,
Thus Love's bright wings are clipped by Time:
    Thus Hope, her soul elate,
Exhales amid this grosser air :
Thus lightest hearts are bowed by care,
    And Genius yields to Fate.

L.A., 1811.

À BALLOON.

In 1830 Messrs. Roger and Green made an
ascent from Kilmarnock. The following anonymous verses were written at the time:—

> Unloose the cords an' let her gang,
> Impatient are the countless throng:
> Some wine an' biscuit in yon pang,
>   And set awa;
> Let's see you in the lift, 'ere lang,
>   Less than a craw.
>
> Hail, Roger, hail! thou fearless chiel,
> The slip'ry wa's o' Heav'n to spiel;
> Beneath yon car the eagle's wheel
>   Their mid-air flight:
> Ye'll supper in the stars, if well,
>   This verra night.
>
> An, now, Sir, as ye're up aboon
> Your grapplin' airns fix in the moon
> Pray fling us thence a parritch spoon
>   Or auld tea-kettle.
> Or ocht that ye can smuggle down,
>   Be't horn or metal.
>
> Tell a' ye either see or hear,
> But no ae sentence less or mair:
> Remember, lad, we'll gar you swear
>   To speak the truth;
> An' if you flinch na' aff the square,
>   We'll quench your drouth.

We fain would ken the cost o' meal,
The weavin', how its paid per ell,
How land does by the acre sell,
  The price o' stock:
An' if for fogin' o' a Bill
  They hang puir folk.

Could there a body get a pie,
Or yill to drink gin he were dry:
We never will the hazard try
  Frae earth to gink.
Unless ye tell what's gaun up bye
  To eat an' drink.

Are poets in the moon respectit,
Or, as they're here, despis'd, neglectit:
Are cottars frae their land ejectit
  When auld and puir:
An' forc'd to beg, fate unexpectit,
  Frae door to door?

What does this planet, sir, appear
When look'd at frae anither sphere:
Is't square, or round, or dark, or clear?
  Some notice tak',
An' gin ye hae an hour to spare
  The shape o't mak'.

Can you distinguish wi' your ee
How far the Poles o't stan' ajee,
Or tell the cause why lately we
  Had sic wat weather?
The North-West passage do you see,
  Or is it blether? *

  * Nonsense.

Ae point learn for our satisfaction,
What is the meanin' o' attraction?
Hae Lunar an' our earth affection
          For ane anither.
An' some strong nat'ral prediliction
          To come thegither?

If Parliament aboon be met,
Try you an' Green to find a seat :
O' ilka motion and debate
          Tak' notes verbatim :
To us they may supply some great
          Desideratum.

Bring Horton plans o' emigration.
To Malthus state the population.
For Sinclair tak' an observation
          O' agriculture :
He'll seize statistic information
          Like ony vulture.

Exert yoursel' an' be na idle,
Nor flee about an' dance or fiddle :
Wi their religion dunna meddle—
          In this be fix'd :
Yet ye may just look gin their Bible,
          Be pure or mix'd.

Tell Lunar bodies something new,
At naething stick to fill them fou,
Ilk plan an' secret frae them screw,
          When drown'd their senses :
Kilmarnock bailies will pay you
          A' just expenses.

In Luna, Sir, what is the practice
O' governments in raisin' taxes?
Just simply tell us what the fact is,
      Since this we spier:
An' if ocht better there the tact is,
      We'll try it here.

What is the uniform o' sogers?
Hae they lang whiskers, like our badgers?
Are whisky-sellers fash'd wi' gaugers,
      Whiles seizin' kegs?
Or kintra women hoax'd by cadgers
      When selling eggs?

Is there the duty aff the leather?
Are besoms made o' broom or heather?
Do wives an' husbands bark at ither
      In hatefu' strife?
How strong, Sir, is the marriage tether?
      Does't haud for life?

Do Kings rule there by right divine,
Or maun the fools a Charter sign?
Their subject's rights to keep in min'
      An' stick by law?
This failing, does the guillotine
      Do ocht ara?

Is Morton* worthy o' our faith,
When he avers, upon his aith,
That, wi' his glass, he sees white claith
      On Luna's hedges?
An' can discern, when clear, plew-graith
      An' swarms o' midges?

* A local optician.

This you maun learn 'bout Luna's markets
Hae we ocht here that peace or war fits?
In kindness, then, sell twa-three forfeits
    O' seeds for Samson:
An' if ye can, some score o'carpets
    For Brown and Tamson.

The poets guid it micht advance
Cou'd ye, in truth, just say at once
That Catrine blunks* wad hae a chance
    To tak' the cad:
Nocht like them can be got from France
    Sae cheap an' guid.

Ye're but a pilgrim in the Moon
Sae bid na there to wear your shoon
Gude sen' you safe an' quickly down,
    Beef-steaks to tak':
We wadna' grudge a white half-crown
    To hear your crack.

      * Cotton goods.

1830.

THE AËRIAL SHIP.

Wonders, sure, will never cease, at least so
    people say,
But always keep on the increase, and very well
    they may;
Since multiplying's all the go, and getting on
    ahead,
We are making wonders now—How shall we
    get our bread?

The aërial ship seems all the go; that's if she'll
    go at all.
Some think she makes a wondrous hit; some
    think she'll make a fall.
'Tis certain if she makes a hit, at the rate she's
    going to go,
To all things standing in her way she'll give a
    sure death blow.

Railways, then, need be no more; balloons no
    more be seen;
The wonders of the great Nassau will then look
    very *green*.
All the shipping may be up, and the seamen in
    despair,
Must sleep on board of boilers, and live on smoke
    and air.

*   *   *   *   *

For though these things may come to pass, they
    have not yet appear'd,
And dangers, when they're out of sight, they
    never should be fear'd.
And all our dread and doubt of this may only
    be a joke,
For projects which we build in air most often
    end in smoke.

James Catnach, 1830.

H

1830.

THE GREAT BALLOON.

What wonders spring up every day, Sirs,
Surely they will never stay, Sirs:
The march of intellect is blooming,
And the mania now is all ballooning.
There's Mr. Green so æromantick
Has built a balloon—in size—gigantic,
Which, when of gas there is a plenty,
Instead of one will *take up twenty*!

The folks now talk both night and noon, Sirs.
Of the wonders of this great balloon, Sirs.

James Catnach.

1840.

THE CLOUDS.

O clouds! ye ancient messengers,
    Old couriers of the sky,
Treading, as in primeval years,
    Your still immensity!
In march how weirdly beautiful
    Along the deep ye tower,
Begirt, as, when from chaos dull,
    Ye loomed in pride and power
      To crown creation's morning hour.

Ye perish not, ye passing clouds!
　But, with the speed of time,
．Ye flit your shadowy shapes, like shrouds
　O'er each emerging clime;
And thus on broad and furlless wings
　Ye float in light along,
Where every jewell'd planet sings
　Its clear, eternal song
　Over the path our friends have gone!

Ye posters of the wakeless air!
　How splendidly ye glide
Down the unfathom'd atmosphere—
　That deep, deep azure tide!
And thus in giant pomp ye go,
　On high and rashless range,
Above earth's gladness and its woe
　Through centuries of change.
　Your destiny, how lone and strange!

Meller, an American writer, 1840.

1850.

ASCENT OF THE PONY NAMED "ROSE," IN MR. GREEN'S BALLOON, FROM VAUXHALL GARDENS. *

Once on a time some years ago
    The famous Mr. Green,
Whom almost all of us have seen
Quitting these grovelling realms below
    For an aërial location,
Some miles above the clouds,
Came to the fixed determination—
That he might better please the crowds
    Who came to gape and stare
    At his exploits in air—
Of carrying with him some animal aloft.
    Should he give a goose an airy dance—
    As once was done in France ?
    No ! Mr. Green
Would make no mockery of the thing,
But would take up a beast of such a size
As might be worthy to be seen by all men's eyes.
    He had a pony.
    The great day came, and all was ready :
Crowds of spectators at Vauxhall were gazing,
And to the spot, Rose from his quiet grazing
    Forthwith was brought.
Into the car by ropes kept steady
    The pony went.

---

* This was in July, 1850, but there was a similar previous
ascent in 1840.

And to prevent his tumbling into the street
A lock was put on each of Rose's feet—
 Or rather, his fetlocks were locked.
Now Mr. Green mounts on the back of Rose,
The ropes are cut, and away in a crack she goes!
Ten minutes more, and this aërial tour was over,
 And at some place in Kent,
 To his unspeakable content,
Our pony was quietly regaling in a field of clover.

(176 *lines in all.*)   *Anonymous.*

18—.

ODE TO MR. GRAHAM, THE AËRONAUT.

Dear Graham, whilst the busy crowd,
The vain, the wealthy, and the proud,
 Their meaner flights pursue,
Let us cast off the foolish ties
That bind us to the earth, and rise
 And take a bird's-eye view!

A few more whiffs of my cigar
And then, in Fancy's airy car
 Have with thee for the skies;
How oft this fragrant smoke uncurl'd
Hath borne me from this little world,
 And all that in it lies!

Away! away! the bubble fills;
Farewell to earth and all its hills!
    We seem to cut the wind!
So high we mount, so swift we go,
The chimney-tops are far below,
    The Eagle's left behind!

Ah me! my brain begins to swim!
The world is growing rather dim,
    The steeples and the trees;
My wife is getting very small!
I cannot see my babe at all!
    The Dollond, if you please.

Farewell, the skies! the clouds! I smell
The lower world! Graham, farewell.
    Man of the silken Moon!

                      Thomas Hood.

(37 *verses in all.*)

               Bright clouds,
Motionless pillars of the brazen heaven,
Their bases on the mountains, their white tops
Shining in the far ether, fire the air
With a reflected radiance, and make turn
The gazer's eye awry.
                      Bryant.

The charm of sky above my head
Is heaven's profoundest azure—an abyss
In which the everlasting stars abide,
And whose soft gloom and boundless depth might
    tempt
The curious eye to look for them by day.

Wordsworth.

Soon shall thy arm, unconquer'd steam! afar
Drag the slow barge, or drive the rapid car;
Or on wide-waving wings expanded, bear
The flying chariot through the fields of air.

Darwin.

"Balloons may be said to resemble babies, insomuch as they are of no use at present, but may become of use in due time."

Dr. Franklin.

"With benevolent wishes each bosom now burns,
  And awe and amazement, both fill it by turns.
  'Where's he going?' cries one, 'Why, he
    shrinks from our sight!
  And where's this poor fellow to quarter to-night?
  If he soars at this rate in his silken balloon
  He'll surely by sunset be up with the moon!'"

THE CLOUD.

The cloud lay low in the heavens,
    Such a little cloud it seemed:
Just lightly touching the sea's broad breast,
Where the rose-light lingered across the west,
Soft and grey as in innocent rest
    While the gold athwart it gleamed.

And or ever the eve was midnight,
    That cloud was lowering black.
Dimming the light of the stars away,
Dimming the flash of the furious spray,
As the breakers crashed in the Northern bay,
    Winds howling on their track.

Anonymous.

WINGS.

Composed by Dolores.   *From the German by Percy Boyd.*

Wings! to bear me over
    Mountain and vale away:
Wings! to bathe my spirit
    In morning's sunny ray.
Wings! that I may hover
    At morn above the sea:
Wings! through life to bear me
    And death triumphantly.

Wings! like youth's fleet movements,
  Which swiftly o'er me pass'd:
Wings! like my early visions,
  Too bright, too fair, to last:
Wings! that I might recall them,
  The lov'd, the lost, the dead:
Wings! that I might fly after
  The past long vanishèd.

Wings! to lift me upward,
  Soaring with eagle flight:
Wings! to waft me heav'nward
  And back in realms of light.
Wings! to be no more wearied,
  Lull'd in eternal rest:
Wings! to be sweetly folded
  Where faith and love are bless'd.

THE FLYING CLOUD.

Cloud! following sunwards o'er the evening sky,
Take thou my soul upon thy folds, and fly
Swifter than light, invisible as air—
    Fly where? ah, where?

Stay where my soul would stay, then melt and
    fall
Like tears at night-time shed, unseen by all,
As some sad spirit had been wandering round
    The garden's bound.

Cloud! sailing westward tinged with purple dye,
Mocking me, as all helplessly I lie :
Ah! cloud! my longing erred, for we were best
      Another rest.

Then lift me with thee to those fields of air
Where Earth grows thin, and upward, upward
  bear,
Till angels meet us with their wings of fire
      That never tire.
               Anonymous.

THE CLOUDS.

I bring fresh showers for the tiny flowers
    From the seas and the streams :
I bear light shades for the leaves when laid .
    In their noon-day dreams.
From my wings are shaken the dews that waken
    The sweet buds every one.
When rocked to rest on their mother's breast
    As she dances about the Sun.
 I wield the flail of the flashing hail,
  And whiten the green plains under :
And then again I dissolve in rain,
  And laugh as I pass in thunder.
               Shelley.

We've sailors now who plough the briny deep,
Through azure skies and rolling clouds they
  sweep:
Invade the planets in an air balloon
And "fright from her propriety" the moon.
                                    Miss Young.

Electric telegraphs, printing, gas,
  Tobacco, *balloons*, and steam,
Are little events that have come to pass
  Since the days of the old régime.
                                    H. S. Leigh.

As men in sleep, though motionless they lie,
Fledg'd by a dream, believe they mount and fly;
So witches some enchanted wand bestride,
And think they through the airy regions ride.
                                    Oldham.

Who quits a world where strong temptations try,
And since 'tis hard to combat, learns to fly!
                                    Goldsmith.

Printed by Eyre and Spottiswoode, East Harding Street, London, E.C.